Making Things Smart

*Easy Embedded JavaScript Programming
for Making Everyday Objects
into Intelligent Machines*

Gordon F. Williams

MAKER MEDIA™
SAN FRANCISCO, CA

Making Things Smart

by Gordon F. Williams

Copyright © 2017 Gordon F. Williams. All rights reserved.

Printed in Canada.

Published by Maker Media, Inc., 1700 Montgomery Street, Suite 240, San Francisco, CA 94111.

Maker Media books may be purchased for educational, business, or sales promotional use. Online editions are also available for most titles (*http://oreilly.com/safari*). For more information, contact O'Reilly Media's institutional sales department: 800-998-9938 or *corporate@oreilly.com*.

Editor: Patrick Di Justo
Production Editor: Melanie Yarbrough
Copyeditor: Kim Cofer
Proofreader: Charles Roumeliotis

Indexer: WordCo Indexing Services
Interior Designer: David Futato
Cover Designer: Karen Montgomery
Illustrator: Rebecca Demarest

July 2017:   First Edition

Revision History for the First Edition
2017-06-29:   First Release

See *http://oreilly.com/catalog/errata.csp?isbn=9781680451894* for release details.

978-1-680-45189-4

[TI]

Table of Contents

PART V.   Putting It All Together

PART VI.   Conclusion

Preface

We humans have endless imaginations. "Making"—whether that's painting, designing, building, or programming—can be one of the most satisfying human experiences. Standing back from a creation and thinking, "I made that," is one of the best feelings I know.

But today, the things we make don't just need to be inanimate objects. We can bring them alive by making them smart. *Making Things Smart* teaches you how to incorporate micro-controllers into intriguing programmable machines.

Using everyday objects and skills, you'll learn how to make a digital camera, a printer, a robot, an early TV, and much more. As you go along you'll learn about the components you're using and the creative history behind them. You'll also learn to code in JavaScript, the popular programming language used by millions of web developers. Because you'll be using a language interpreter you'll able to build up your sofware line by line and see the effect of each bit of code as you add it.

I love making and have done it all my life. I hope this book inspires you to create and learn, and have fun along the way.

Conventions Used in This Book

The following typographical conventions are used in this book:

Italic

> Indicates menu items, new terms, URLs, email addresses, filenames, and file extensions.

`Constant width`

> Used for program listings, as well as within paragraphs to refer to program elements such as variable or function names, data types, statements, and keywords.

`Constant width bold`

Shows commands or other text that should be typed literally by the user.

`Constant width italic`

Shows text that should be replaced with user-supplied values or by values determined by context.

This element signifies a tip or general note.

This element indicates a warning or caution.

Using Code Examples

Supplemental material (code examples, exercises, etc.) is available for download at *https://github.com/espruino/making-things-smart*.

This book is here to help you get your job done. In general, if example code is offered with this book, you may use it in your programs and documentation. You do not need to contact us for permission unless you're reproducing a significant portion of the code. For example, writing a program that uses several chunks of code from this book does not require permission. Selling or distributing a CD-ROM of examples from Make: books does require permission. Answering a question by citing this book and quoting example code does not require permission. Incorporating a significant amount of example code from this book into your product's documentation does require permission.

We appreciate, but do not require, attribution. An attribution usually includes the title, author, publisher, and ISBN. For example: "*Making Things Smart* by Gordon F. Williams (O'Reilly). Copyright 2017 Gordon F. Williams, 978-1-680-45189-4."

If you feel your use of code examples falls outside fair use or the permission given above, feel free to contact us at *bookpermissions@makermedia.com*.

O'Reilly Safari

Safari (formerly Safari Books Online) is membership-based training and reference platform for enterprise, government, educators, and individuals.

Members have access to thousands of books, training videos, Learning Paths, interactive tutorials, and curated playlists from over 250 publishers, including O'Reilly Media, Harvard Business Review, Prentice Hall Professional, Addison-Wesley Professional, Microsoft Press, Sams, Que, Peachpit Press, Adobe, Focal Press, Cisco Press, John Wiley & Sons, Syngress, Morgan Kaufmann, IBM Redbooks, Packt, Adobe Press, FT Press, Apress, Manning, New Riders, McGraw-Hill, Jones & Bartlett, and Course Technology, among others.

For more information, please visit http://oreilly.com/safari.

How to Contact Us

Please address comments and questions concerning this book to the publisher:

Make:
1160 Battery Street East, Suite 125
San Francisco, CA 94111
877-306-6253 (in the United States or Canada)
707-639-1355 (international or local)

We have a web page for this book, where we list errata, examples, and any additional information. You can access this page at *http://bit.ly/making-things-smart*.

Make: unites, inspires, informs, and entertains a growing community of resourceful people who undertake amazing projects in their backyards, basements, and garages. Make: celebrates your right to tweak, hack, and bend any technology to your will. The Make: audience continues to be a growing culture and community that believes in bettering ourselves, our environment, our educational system—our entire world. This is much more than an audience, it's a worldwide movement that Make: is leading. We call it the Maker Movement.

For more information about Make:, visit us online:

Make: magazine: *http://makezine.com/magazine*
Maker Faire: *http://makerfaire.com*
Makezine.com: *http://makezine.com*
Maker Shed: *http://makershed.com*

To comment or ask technical questions about this book, send email to *bookquestions@oreilly.com* or visit *http://forum.espruino.com*.

Acknowledgments

I'd like to thank the team at Maker Media for giving me a chance with *Making Things Smart* —despite it being my first book—and for their patience as I have come to grips with the process of book writing! I'd also like to thank Brian Jepson and Anna Kaziunas France, who are no longer at Maker Media but were instrumental in getting me started.

My wife Marianne has been amazing—not just for her help with this book and my work in general, but for giving me the confidence to start working for myself almost 10 years ago, and for her continued support of my crazy ideas since then!

This book and Espruino in general wouldn't have existed without the amazing support of my KickStarter backers. Their initial backing helped make the first Espruino board, and since then their continued support and enthusiasm has helped Espruino go from strength to strength. Members of the Espruino forum have been hugely helpful, and more recently my Patreon supporters have meant I can dedicate more time to working on more exciting Espruino projects. I'm also hugely grateful for the thoughtful bug reports and improvements that come through GitHub.

I now use open source software almost exclusively, and depend on tools like Linux (Mint), GCC, Chromium, Eclipse, Atom, Gimp, Inkscape, and LibreOffice. GCC often seems overlooked, but when I started work on Espruino, many embedded devices didn't have access to a rock solid, free C compiler.

Laurent Desseignes and Sebastien Marsanne at ST Microelectronics have been hugely supportive with the STM32 Espruino boards, and Michael Dietz and many other developers at Nordic Semiconductor have provided a huge amount of help during the development of Puck.js.

François Beaufort of Google has gone well out of his way to help with Web Bluetooth and Physical Web support in Chrome, and David Park of Green Park Software brought Web Bluetooth to iOS with the WebBLE app. I'd also like to thank Rob Moran, Jonathan Austin, Hugo Vincent, and Simon Ford at ARM mbed, who have given me invaluable advice as well as helped to pull some strings to help publicize Espruino and ensure that things like the micro:bit port of Espruino became a reality.

The Cambridge University Computer Lab has been a great help—not just for my education, but for their continued support even after graduation. Stuart Newstead especially has been a fantastic mentor, and has helped make sure I didn't spend all my time writing code at the expense of everything else!

I'd also like to thank Tim Hunkin and Rex Garrod, whose "The Secret Life of Machines" TV series from the 1980s was a huge influence; the printer project in this book bears more than a passing resemblance to their re-creation of a fax machine.

Finally, I would never be doing this if it wasn't for my parents, Fred and Pat Williams. Without their help and support (and an endless supply of computer gear and electronic components) growing up, I'd have never got into software and electronics. I spent my childhood making and experimenting, and the TV project in particular is directly based on a device made with my father. I hope that in some way this book helps a few more children experience the sense of excitement and wonder that I did.

Introduction | 1

Why Make Things Smart?

It's easy to categorize different forms of creativity and think that they don't really overlap. We differentiate makers in our minds in the way we describe them: artists, computer programmers, architects, or fashion designers.

These different craftspeople are, however, united by the fact that over the last few years they have all been incorporating smart devices into their creations. We've seen an explosion of home-automation, wearable technology, and intelligent art installations. Computer programs are no longer just in our laptops; we have programmable kettles, clothes, cars, and climate controls.

Even within the sphere of technology, distinctions are drawn between mechanics, electronics, and software. Each is considered to be a very separate skill, and very few people combine them. Look online and you'll see projects with beautiful woodworking skills, complex electronics, and ingenious computer software. However, it's rare to see all three together.

But that distinction isn't helpful for smart things, which need to exist in the real world. In this book I'll show you how to combine simple hardware, circuits, and sofware to make intriguing machines using everyday materials and basic components.

Once you've explored the projects in this book, you will be able to start inventing and using your new skills and expertise when designing smart things, whether that means you make beautiful moving sculptures, functional data recorders, elegant luminescent handbags, or gadgets to automate your home.

Learning through Making

The writer George Kneller said that "It seems to be one of the paradoxes of creativity that in order to think originally, we must familiarize ourselves with the ideas of others."

Let's face it, you probably already have a better printer, digital camera, and TV than the ones you'll make with this book. But the chances are that you won't really know how they work and certainly wouldn't think that they were things you could make for yourself at home with basic components and everyday materials.

In a world where most people have a phone that acts as a camera, video recorder, music player, browser, and GPS in one it can be daunting to try to understand what is actually going on in the technology we use. *Making Things Smart* takes you back to the basics. As you make the projects from scratch you'll learn the principles from which much more complex technology derives and gain an understanding of the fundamental building blocks on which so much of our modern world is built.

As you engage with the projects and instructions in this book you'll also develop your own skills and knowledge. With that experience you'll be able to bring your ideas to life and start creating your own smart things.

Making in JavaScript

Making Things Smart teaches you to program your hardware in JavaScript using a language interpreter called Espruino, which runs on your microcontroller. JavaScript is one of the most popular programming languages on the internet, with lots of online support and resources. I developed the Espruino interpreter to bring JavaScript into the world of smart devices because the ease with which you can change code encourages iterative development.

In many books on microcontrollers you'll find full listings of computer code, but *Making Things Smart* is different. By programming using Espruino you can build software up line by line, seeing the effect of each bit of code as you add it. Espruino doesn't just "crash." If your code produces an error, you get a helpful message explaining what went wrong and pointing to where in your code it happened.

Not only is this great for learning, but it's a lot more fun too, and hopefully will give you the confidence to experiment yourself as you go along. Full listings (*https://github.com/espruino/making-things-smart*) are still available online if you just want to try out the finished article!

Tools and Materials

The projects in this book are designed to be made with very simple materials and tools. Most people will have the following items laying around.

There's quite a bit of cardboard cutting, so you'll need some relatively heavy-duty scissors.

Sometimes you'll need to cut things that won't be possible with scissors, and a craft knife (of any type) and some kind of surface to cut on are required.

Occasionally you'll need to screw a few things into blocks of wood, or attach servo plates onto servo motors. Any standard screwdrivers will do.

Sometimes you'll need to bend paperclips or wire at sharp angles, and a pair of thin-nose pliers will make your life a lot easier.

You'll need to be able to cut and strip wire. I'd strongly recommend the *T-Stripper* from *Ideal* (or similar-looking tools) over fancy-looking automatic wire strippers. If you have a choice, get one that will take 24 AWG wire or thinner (thinner wire has a larger AWG number).

There are a few nails that need hammering in, so you'll need a hammer. However, a very light-duty hammer is absolutely fine.

For all projects apart from the last one you won't need a drill, but you might find a hand drill makes things easier. For the last chapter if you have a drill press (or just a way of making straight holes) you'll get better results, though!

Finally, you'll need a hot glue gun. Quite a few of the projects use hot glue to fasten things together as it's relatively quick to set, sticks well to cardboard, and can be peeled off most things if you position them wrong or want to reclaim them!

PART I

Microcontrollers

Most of us interact with tens or even hundreds of microcontrollers every day without even realizing it.

They're the perfect computer, making our life easier without ever breaking and making it difficult.

In these chapters we'll look at what they are, and how to get started with them using Espruino and JavaScript.

What Is a Microcontroller? | 2

A microcontroller is a small, self-contained computer. Your PC, and possibly your phone, might have many discrete components—RAM, nonvolatile memory like hard disks or SSDs, oscillators, and power supplies. Most microcontrollers, on the other hand, have everything they need on one piece of silicon: RAM, flash memory, oscillators, and even voltage regulators (if needed). You can get many microcontrollers to work just by connecting them straight to a battery of the correct voltage.

The two types of processor are getting increasingly blurred. Microcontrollers can now be faster than desktop computers were 20 years ago, and the SoC (System on Chip) processors in devices like mobile phones have more and more components integrated inside them to reduce costs. The real difference is in the intended usage. Microcontrollers are designed to be embedded into things and to do just one task (often without a display); normal computers are meant to be more general-purpose.

Most of us have cursed our computers for not doing what we expected, but microcontrollers are computers that work so well and so reliably that much of the time we're not even aware that they exist. Every day, you probably interact with 100 or more microcontrollers: in your watch, phone, keyfob, car, wireless credit card, bike lights, and so on. They work silently in the background, making your life just that little bit easier.

In 2015, Advanced RISC Machines (ARM) licensed a staggering 15 billion ARM cores. That's two for every person on the planet. ARM microcontrollers are only one of many different types of microcontroller available, so the actual number of microcontrollers produced is even higher. They're literally everywhere.

Microcontrollers come in many shapes and sizes: from as small as a grain of sand, to as large as a postage stamp. They can come with as few as 16 bytes of RAM, or a million. They also come in many different architectures (the instructions they execute and how they do it).

Common architectures for microcontrollers are PIC and MIPS (used by Microchip), AVR (used by Atmel), and ARM (used by ST, Atmel, Nordic, Freescale, Silicon Labs, and many others). PIC and AVR started life as 8-bit processors. For these, each single instruction operates on 8-bit values (numbers between 0 and 255). This makes the processors very small and efficient at simple tasks, but more complex problems such as multiplying a 32-bit number can end up taking much longer.

The first PIC and AVR microcontrollers shipped with very small amounts of memory, for instance the AT90S8515 (an AVR microcontroller) had 8kB of flash memory and 512 bytes of RAM. Only being able to use 8-bit numbers to access the RAM wasn't a big problem as you could easily access one half, and then the other. These processors also used Harvard architectures, where the RAM and the flash memory are completely separate. It made a lot of sense as the design could be kept very simple: you just got instructions for the microcontroller to execute from flash, and then stored data in RAM.

However, as the memory available in microcontrollers has increased (along with the expectations of programmers who don't want to worry about whether their data is in flash or RAM), the processor has to spend significantly more time doing calculations just to access the correct part of memory. The compromises that made sense for 0.5kB of memory start to look extremely inefficient in larger systems.

ARM History

The first ARM chip was designed in 1985 by Steve Furber and Sophie Wilson at Acorn Computers in Cambridge, England. At this point Acorn was just starting to ship its BBC Micro Model B+128, which had (as the name suggests) 128kB of RAM. That is 64kB more memory than its 16-bit processor could directly address.

The engineers at Acorn were obviously looking forward and knew that they'd be building computers with even more memory in the next few years. They wanted a way to easily access more memory efficiently, so they decided to make a 32-bit processor.

Why not a different number of bits?

Pretty much every computer uses multiples of 8 bits, so the obvious choice above 16 bits would have been 24 bits. However, this would have meant that memory accesses would have needed to multiply or (crucially) divide by 3. If you want to access the 18th byte, that would be in the 18 / 3 = 6th *word*.

In binary math (as in normal math), dividing by most numbers is pretty hard. However, if you're dividing by your base of arithmetic then it's easy.

For instance, we use base 10, so if you wanted to divide 3732867532 by 10 it's easy; you just take the 2 off the end. Dividing by 100 or 1000 is just as easy as well. If you wanted to divide by 7 it's a lot more painful!

Dividing works just the same in binary, which is base 2. If you want to divide 10010101110 by 3 it's hard, but dividing by 2, 4, 8, 16, etc., is easy: you just take digits off the end.

For the folks at ARM, the obvious choice was a 32-bit processor, so they could just multiply and divide by 4 for addresses. For modern computers, where you often have more than 4GB of RAM, they wanted to increase the size again. This time they had to multiply and divide by 8, so they moved to 64 bits, which will reference enough data to keep us going for quite a while!

The ARM core was designed from the start to be a 32-bit processor, with each instruction dealing with numbers between 0 and 4294967295. While this added some complexity, it wasn't as bad as you might think. The design of each bit of register and arithmetic logic unit was very modular, so once there was a working design for one bit, it could just be repeated 32 times.

Having the ability to store and work on 32-bit numbers at once meant that the ARM could easily have a von Neumann architecture, where both the instructions and data are stored in the same address space. A single instruction could load data from RAM or ROM, and the address itself determined which area it was. It made the ARM processor's instruction set very simple, and so very easy to write code for.

The ARM core was originally designed for fully fledged computers. It's still used as the primary processor in a few of them today, as well as almost all mobile phones and tablet computers. However, as microcontrollers have become more powerful, the ARM architecture has found a home in those as well.

Programming a Microcontroller

Computers read instructions from memory that tell them what to do. To run quickly and efficiently, these instructions need to be easy for the computer to understand, so a lot of effort goes into their design. Unfortunately, what is easy for a computer to understand often isn't easy for a human!

As an example, you might want to add the numbers from 1 to 10 together. The actual instructions for the ARM (the *machine code*) might look like this:

```
e3 a0 50 00
e3 a0 40 01
ea 00 00 01
e0 85 50 04
e2 84 40 01
e3 54 00 0a
da ff ff fb
```

Obviously this isn't going to be much fun for a human to understand or write. The processor is 32 bits, so each line of 4 bytes represents an instruction. If we could write what each one did it would be a lot easier to understand:

```
e3 a0 50 00          mov   r5, #0
e3 a0 40 01          mov   r4, #1
              loopstart:
ea 00 00 01            b    loopcheck
e0 85 50 04          add   r5, r5, r4
```

```
e2 84 40 01            add   r4, r4, #1
          loopcheck:
e3 54 00 0a            cmp   r4, #10
da ff ff fb            ble   loopstart
```

This is what we call assembly language. It's a textual representation of the actual instructions that the computer executes. Each line (apart from the labels that end with a colon) represents one instruction. It's covered in more detail in Appendix B where you can try writing some assembly with Espruino.

Originally, the process of getting from assembly language to machine code was a slow, manual process, but now there's some software called an assembler that will convert it automatically. However, because our microcontroller doesn't have any software on it yet, we need to run the assembler on our PC first, get the machine code it creates, and then send that to the microcontroller.

To be fair, writing assembly code is pretty hard too. It would be better if we could write code in a language that was better suited to humans. The following code is written in a programming language called C. It's called the *source code*:

```
int a = 0;
for (int b = 1; b <= 10; b = b + 1)
  a = a + b;
```

Some software called a compiler (just a more complicated version of the assembler) can run on your PC, and will take this simpler code and convert it into machine code that you can send to the microcontroller. It's called *cross compilation* (because you're using one type of computer to compile code for another).

Optimizing Compilers

In reality, the assembly code shown here would never be generated. Modern compilers usually *optimize* the code you write. For instance, if you wrote 1+2 then the compiler would just insert 3 into the code.

Going even further, a modern compiler would realize that the previous loop always gives the same result (55) and would insert that instead. It will even reorder your code and remove parts of it that it can deduce will never be executed.

To aid with debugging you can often disable these optimizations, so you can manually step through every statement in order. That's how I got the original assembly code (see preceding code).

This is the way most microcontrollers are programmed, but there's a problem. The code shown previously bears no real resemblance to the machine code that is sent to the microcontroller. If the microcontroller has a problem executing some instruction (let's say we want to sum the numbers from 1 to 100000, and the number gets so big it can no longer fit in 32 bits), we don't have any obvious way to get back from the machine code to find the original source code that was at fault.

In most professional tools, the problem is solved. You can connect a special piece of hardware to the microcontroller that will allow you to see exactly what is happening and map it back to your original code. However, for a lot of people (for instance, those using normal Arduino boards), the microcontroller is essentially a black box.

Once the source code is compiled and sent to the microcontroller, there is no *feedback*. The microcontroller runs your code and does exactly what you told it, but if there is an error the microcontroller is unable to tell you where that error occurred. If you want to see what your code is doing, you have to explicitly add code that will tell you what's going on, either by changing the voltage on a pin, or by sending characters through an on-chip peripheral that you can then read back on your computer.

One solution to this is to have some special machine code on your microcontroller that will read your source code and execute it directly. This is called an *interpreter*. It isn't as efficient, because now the microcontroller has to execute the interpreter code as well as your code whenever it runs. However, the interpreter can check your code as it runs, and can report any errors when they occur rather than just crashing.

JIT—Just in Time Compilation

Many modern interpreters have something called JIT compilation. They actually contain a compiler, and take your source code and compile it to machine code as it is about to be executed.

This means that once your code is compiled, the interpreter no longer has to interpret each instruction and your program can run quickly. Compilation often takes a while, though, so most interpreters only choose to compile your code if they've discovered that it is being used a lot. If the code you've written is only called once, then they will only interpret it rather than compiling.

Google's V8 interpreter is a notable exception. It *always* compiles your code, so is very fast. If you use your code a lot, it then tries again and spends even more time trying to find ways to optimize it to make it run more quickly.

Having an interpreter on your microcontroller also means that you don't need a compiler installed on your host computer. Pretty much everything you need is now on the microcontroller itself. The only thing you need on your host computer is a way to send and receive characters.

That's how Espruino works: some code that is already on the microcontroller allows it to execute the JavaScript code that you type, without requiring you to have any special software installed on your PC.

Getting Started with Espruino

Now it's time to get started with some hardware!

For the sake of simplicity we're going to cover just the Espruino Pico, shown in Figure 3-1 (which is very similar to other Espruino boards). It is possible to run the Espruino firmware on other boards (once you have loaded it on), but we won't cover that here. Check out the Espruino website (*http://www.espruino.com*) for more detailed information on supported boards and installation instructions.

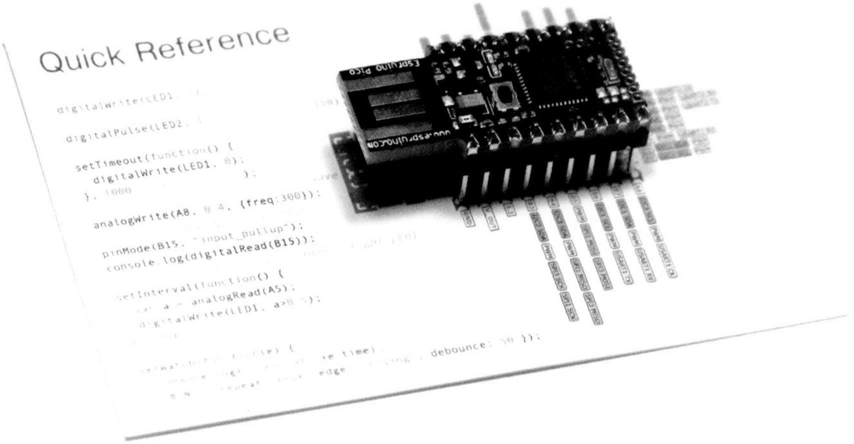

Figure 3-1 *The Espruino Pico board and its quick reference card*

The Espruino Pico is available from a range of distributors across the world (*http://www.espruino.com/Order*).

If you're using Puck.js (a self-contained Bluetooth device that runs Espruino), you don't have to use USB, so you don't need to follow the first part of this chapter.

Instead, just head over to http://www.puck-js.com/go *and follow the instructions there.*

Getting Ready

When connected to your PC via USB, the Espruino Pico should appear as a standard USB *Virtual COM Port* peripheral. Unfortunately, on some platforms there are a few things you need to do to get this to happen.

*Don't fancy typing these links? You'll be able to find them in the Espruino Quick Start guide (*http://s.espruino.com*).*

Mac and Chromebook

There's nothing to do. You can head straight to plugging the Pico in.

Windows

If you have any version of Windows other than XP (10, 8, 7, Vista), you'll need to download ST's Virtual COM Port drivers version 1.4.0 (*http://www.espruino.com/files/stm32_vcp_1.4.0.zip*).

If you have Windows XP, you'll need version 1.3.1 (*http://www.espruino.com/files/stm32_vcp_1.3.1.zip*).

Once you have downloaded the file you'll need to:

1. Open the ZIP file.

2. Run the executable inside (there's no need to extract any other files).

3. Go through the installer's steps.

4. Open Windows Explorer and navigate to *C:\Program Files (x86)\STMicroelectronics \Software\Virtual comport driver* (or just *Program Files* on 32-bit systems).

5. Run the executable for your system (*amd64* for 64 bit, or *x86* for 32 bit).

Linux (Including Raspberry Pi)

On Linux the Espruino Pico will *just work*, but by default normal users won't have permission to access it. To fix this:

1. Download the file *45-espruino.rules* (*http://bit.ly/2pl85k2*).
2. Copy it to */etc/udev/rules.d* with `sudo cp 45-espruino.rules /etc/udev/rules.d`.
3. Run `sudo udevadm control --reload-rules` to reload the udev rules without having to restart.
4. Type `groups` and make sure your user is a member of the `plugdev` group.
5. If not, type `sudo adduser $USER plugdev` and then log out and back in.

Plugging In

For most of the projects in this book you'll want a USB Type A to Type A extension lead (that's the one with a male version of the large rectangular plug on one end, and a female version of the same plug on the other). If you're a user of a modern Mac, you may also need to get a dongle that will provide you with a standard USB Type A socket.

You can plug the Pico straight into the side of your computer, but you'll then find it harder to plug other things into the Pico. Having a USB extension lead gives you a little more flexibility.

 Other Espruino boards like the Original Espruino and Espruino WiFi use a Micro USB connector (the one used on the majority of non–high-end mobile phones as of 2017).

The only trick to plugging the Espruino Pico in is to do it the right way around (shown in Figure 3-2). You want the gold contacts on the Espruino Pico to be facing the plastic insert in the USB plug, not the metal shield (otherwise they won't make contact).

If all goes well, you should see the red LED on the Espruino board flash for a fraction of a second. On Windows it could take several minutes for the board to be recognized as a communications port, but you can get started installing the Espruino IDE while that happens.

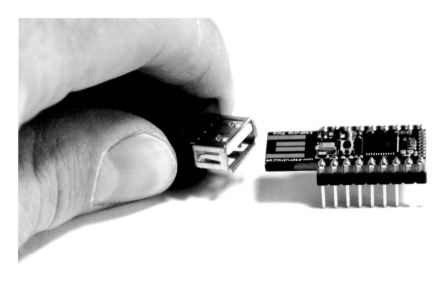

Figure 3-2 *Plugging Espruino into USB—ensure the plastic part of the USB connector is facing toward the gold USB contacts on the Pico*

Installing Software

Espruino should now be available to use with any terminal application (it appears as a normal communication port); however, to get the best experience with a nice editor and debugging, it's best to install the Espruino Web IDE.

*At the moment, the Web IDE is available as a handy application that can be installed from inside the Chrome web browser (**Figure 3-3**). However, Google has announced that it will be shutting down the Chrome Web Store at the end of 2017.*

*This means the Web IDE (**http://www.espruino.com/Web+IDE**) will be made available as a separate download.*

Figure 3-3 *The Chrome web browser's logo*

First, you need to get the Chrome web browser (*http://google.com/chrome*) (Figure 3-4).

Get a fast, free web browser

Figure 3-4 *The Chrome web browser download page*

Once it's installed, open Chrome and head over to the *Chrome Web Store* (Figure 3-5). Type `Espruino` into the search box and click the *Espruino Web IDE* item.

Figure 3-5 *The Espruino Web IDE's entry on the Chrome Web Store*

 If you can't find it, you can go directly to the app by navigating to http:// bit.ly/2ojgtBl.

Now just click the ADD TO CHROME icon in the top right of the app to install it.

You can now click ↑ LAUNCH APP to launch the Web IDE. It'll also be available on Chrome's Home Screen or App Launcher. You should get something like the screen shown in Figure 3-6.

Figure 3-6 *The Espruino Web IDE*

Connecting

When you first start you'll see a quick onscreen tutorial to guide you through the different parts of the IDE.

Now to get started! Click the orange icon in the top left of the window, and you should see a pop-up window showing a list of available ports. One of them should say `Espruino` (see Figure 3-7).

Figure 3-7 *The Web IDE's Port Selector showing a connected Espruino board*

 If you don't see anything, check out http://www.espruino.com/Troubleshooting *for some ideas of what might be wrong and how to fix it.*

Now, click the menu item for the Espruino board. After a second or two the menu will disappear and the screen will show Connected on the left. It'll look like Figure 3-8.

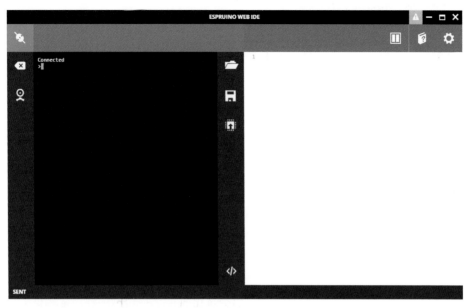

Figure 3-8 *The Web IDE once connected to an Espruino board*

Updating Firmware

Most likely, when you connect you'll see a yellow warning symbol in the top right of the window (as in Figure 3-8). This tells you that the firmware on Espruino is out of date. Espruino's software gets changes and improvements all the time, so the version that came on it from the factory will almost certainly be out of date.

To avoid any unwanted problems it's a really good idea to update. To do this:

1. Click the yellow warning triangle (Figure 3-9). You can also get to the window by clicking the settings icon in the top right, followed by *Flasher* on the left of the *Settings menu* that pops up.

Figure 3-9 *The Web IDE's menu for updating the firmware on an Espruino device*

2. Now click the *Flash Firmware* button and follow the instructions. Updating the firmware should only take about a minute.

3. Check that the lights aren't pulsing on and off on your Espruino board. The pulsing tells you that the board is still in bootloader mode. If it is, just unplug it and plug it back in.

4. Now click the connection icon in the top left again to reconnect to the board, then choose it from the list.

First Commands

Finally you're ready to go!

1. Click in the big dark area on the lefthand side of the IDE.

   This is the Espruino Console. When you type on the left of the IDE you are communicating directly with the Espruino board.

2. Now type `1+2` then press `Enter`.

   Espruino will display `=3`. It has interpreted your formula in the board itself and returned the result.

   We now want to type `digitalWrite(LED1,1)`. This will execute the function called `digitalWrite` (see Chapter 4) that will put `LED1` into the *on* state, turning the red LED light on. We can save some typing using the autocomplete feature.

3. Just type `di` then press `Tab`.

Espruino will autofill the common characters it knows about (`digital`) and will then prompt you with possible options, like `digitalWrite`, `digitalRead`, and `digitalPulse`.

4. Type `W` then press `Tab` again (it's important that `W` is uppercase as JavaScript is case sensitive).

   Espruino will now autofill the rest of `digitalWrite`.

5. Now type the rest of the command—**(LED1,1)**—and then press `Enter`.

   The red LED should now have lit up!

   Espruino will write `=undefined`. This is because unlike `1+2`, the `digitalWrite` function didn't return a value, and functions that don't return a value actually return the special JavaScript type `undefined`.

   What if we want to turn the LED off? We just want to send `0` and not `1`. You can use the command history to go back and edit the line.

6. Press `↑`, and `digitalWrite(LED1,1)` should now be displayed.

7. Use `←` to move the cursor back to just after `1`, use `Backspace` to delete it, and type `0`.

8. Now press `End` to move the cursor to the end of the line, or use `→`. If you don't do this and then you press `Enter`, you'll end up putting a new line in the middle of the statement!

9. Now press `Enter` to send the `digitalWrite(LED1,0)` line.

   The LED should now be off.

10. Enter **digitalRead(BTN)** to read the state of the button. Espruino should return `=false` because the button isn't pressed.

11. If you press the button on the Pico, and then press `↑` and `Enter` to rerun the previous `digitalRead` command, you will get `=true` returned.

You've now played around on the lefthand side of the IDE. This is Espruino's *console* (sometimes known as a REPL). It's a direct connection to the microcontroller itself. When you press a key on your keyboard, that key is converted to bytes of data that are sent to the board, and the board sends data back for each character.

You can get the same effect using any VT100-style terminal application (for example, *screen* on Mac or *PuTTY* on Windows), but the Web IDE is capable of a lot more.

The Editor

On the righthand side of the IDE is the JavaScript editor. It should have come prefilled with some code like this:

```
var  on = false;
setInterval(function() {
   on = !on;
   LED1.write(on);
}, 500);
```

This code flashes the red LED on and off. Rather than use `digitalWrite(...)` as we did, it uses `LED1.write(...)`. It's just another way of doing the same thing (setting the state of a digital output). `setInterval` calls the function supplied to it every 500 milliseconds, so twice a second. For more information on what this all means, check out Chapter 4.

The editor displays different words in different colors, to make the text easier to read. It's also got some other nice features:

1. Click a word like `write` and press `Ctrl`-I. A pop-up window will appear that explains what the function actually does.

2. At the end of the code, type `di` and press `Ctrl`-Spacebar. Much like autocomplete on the lefthand side of the IDE, you'll get a list of functions, this time with descriptions for each. Delete what you typed, so we can upload the original code!

You can also search, find definitions, and rename variables. Go to *Settings* (the icon in the top right), *About*, and scroll down for more information.

But for now, we just want to get the code working, so click the *Upload* button (⬛). If you hover your mouse over the button, it'll say *Send to Espruino*. This will upload the code on the right into Espruino and make the LED flash. It does some important things in the process:

• First, it resets Espruino. This is like typing `reset()` on the lefthand side of the IDE.

• Then, it looks at the code you've written and if you're using any modules with `require` (we'll do this in Chapter 15) it will automatically download them to the board.

• It also looks for code that needs compiling or assembling (see Appendix B), and also adds line numbers for debugging, among other things.

All these features mean that often it's far better to build your programs on the righthand side of the IDE (it's a lot easier to load and save your code from there too).

The lefthand side is most useful for interacting with your program once it has been uploaded, for example, for looking at variables or tweaking functions.

For instance, now that you've uploaded the program on the righthand side, you can look at how it's working.

1. Type `on` then press `Enter`.

   This will print the value of the variable `on`. It'll be `true` or `false`.

2. Press ⬆ and Enter to evaluate `on` again, and see if it changes. You may have to do this a few times, but the state of `on` will correspond to the state of the LED at the time that you pressed Enter.

3. Type `changeInterval(1, 200)`. This will change the speed of the first interval created by `setInterval` to be 200ms, and the light will flash a bit faster.

4. Now, type `dump()` and press Enter.

Espruino will write something that looks a bit like this:

```
var on = false;
setInterval(function () {
  on = !on;
  LED1.write(on);
}, 199.99980926513);
```

It's not exactly your code, but it's close. Espruino has reconstructed the *current state* of the program in a human-readable form, based on its internal data structures. If you make changes on the lefthand side of the IDE and want to know what they were, this is what you'll have to do.

Finally, what if we want to save what we've done, so it happens even after we unplug power and reapply it?

1. Simple. Just type `save()` then press Enter. Espruino will save your code into flash memory.

2. Click the *Connect/Disconnect* button in the top left and then unplug Espruino.

3. Now plug Espruino back in. The LED should start flashing!

If you plug Espruino into any source of power, it will now start up and the LED will keep flashing.

But let's face it, this is going to get annoying really quickly! Let's get rid of it:

1. Reconnect the Web IDE.

2. On the lefthand side, type `reset()`. This will completely reset Espruino, stopping the flashing.

   This is only temporary though! If you disconnected, unplugged Espruino, and plugged it back in, the light would keep flashing.

3. Type `save()` again. This will save the current (freshly reset) state back into Espruino's flash memory, meaning that the next time you plug it in, there's no annoying flashing.

This works in most cases, but occasionally you will be able to get Espruino into a state where you are unable to type `reset()`*.*

If this happens to you, just follow the steps for re-flashing the Espruino firmware, and everything will be wiped back to factory settings.

On the whole, we'll use the lefthand side of the IDE (the console) a lot in this book, mainly so you can build up functionality layer by layer, without having to restart your program each time. However, for more complex projects you'll almost certainly find that it is easier to start using the righthand side.

And now, let's delve a bit deeper into JavaScript!

Getting Started with JavaScript | 4.

JavaScript is an amazingly powerful language, but it has a few quirks! This chapter will take you on a quick tour of JavaScript's features. We'll assume you know a little bit about programming (maybe in a different language), and will explore what is different about JavaScript.

Because JavaScript is an interpreted language, you have to use an interpreter to run the JavaScript code that you have written. In this chapter and throughout this book we'll be using the Espruino JavaScript interpreter. However, all of the examples and code in this chapter will work in the same way using most other JavaScript interpreters, such as the V8 interpreter that's built into the Google Chrome web browser.

 ECMAScript 6, known as ES6 or ECMAScript 2015, is a new version of JavaScript released in 2015. We'll refer to it as ES6.

At the time of writing this book ES6 still doesn't have full support in all web browsers (especially on mobile platforms), and not all of its features are implemented in Espruino. We won't use those features in our examples or talk about them in detail in this chapter, but a few of the more exciting features are mentioned throughout this chapter.

Getting Ready

We'll use Espruino to start coding in JavaScript, so you need to have your board set up as you had it in the previous chapter:

1. Plug the Espruino board into your computer.
2. Run the Espruino Web IDE.

3. Click *Connect* in the top left of the screen and select the Espruino board from the list.

Once connected you'll need to type commands on the lefthand side of the IDE. This is the Espruino Console, where any commands you type are executed as soon as you have typed them and pressed `Enter`.

Comments

In JavaScript, comments can either be short and included between code on one line or can run to the end of the line. The two different kinds of comments look like this:

```
some code here /* This is a short comment */ some code here
```

```
some code here // This is a comment until the end of the line
```

We'll use these two kinds of comments throughout this book. The purpose of comments is simply to help explain the surrounding code, so the interpreter will just ignore them. This means that if you're copying out the code by hand, rather than loading it off the internet, there's no need to copy the comments.

Data Types

When you're writing code, you'll need to define the data that you want to work with.

Undefined

The most basic data type in JavaScript is `undefined`. You'll see this a lot: it basically means *nothing*. For example, if you try to access some data that's not there, or if you look at the result of a function that doesn't return anything, you'll see the output `undefined`.

Numbers

Probably the next simplest type of data is a number.

Type `1234` into the lefthand side of the Espruino Web IDE and press `Enter`.

This will be executed by Espruino and `1234` will be displayed.

If you type `12.34`, it will be executed in the same way. It defines a fractional, or floating-point number.

Strings

Strings are the next kind of data to think about. These are just sequences of characters that you will need to define if you want to work with text.

Type `"Hello"` and press `Enter`.

A simple `="Hello"` will be displayed, because that's the value of what we wrote.

In this example, the simple string starts and ends with a `"` character. But a simple string can also be started and ended with a `'` character.

Remember this, because simple strings will become useful later. Because the second `"` or `'` character *ends* the string, you have to be careful about using these characters in a string directly. However, there are ways to work around this.

One option is to use the escape character, `\`. The escape character tells the interpreter that the next character does not start or end the string. For example, `"Hello \"quoted\""` represents the characters `Hello "quoted"`.

The other option is to define the string using whichever of the `'` or `"` characters you have not used to start the string. For example, `'Hello "quoted"'` also represents the characters `Hello "quoted"`.

Booleans

Ok, let's move on to Booleans. A Boolean is a value that only has the value `true` or `false`.

You can define a Boolean by entering `true` or `false`. Alternatively, you can create a Boolean by comparing numbers.

For example, type `1 < 2`, then press `Enter`.

This will return the Boolean `true`, because 1 really is smaller than 2. Similarly, if you type `2 < 1` the Boolean returned will be `false`.

You might be tempted at this point to try typing `1 = 2` *but this won't work*.

This is because in JavaScript (and many modern programming languages) `=` is used to assign one value to another and not to compare two values. Instead, if you want to compare two values to see if they are the same you have to use two equals characters, like `==`.

So, if you type `1 == 2` JavaScript will return the Boolean `false` because 1 is not equal to 2.

Math

So far, we've looked at how to define and compare values. To do anything constructive with our software, we'll need to be able to perform operations on the values we defined, and that's where math operators come in.

Math with Numbers

Now we'll try using JavaScript to do some math:

1. Type `1 + 1`, then press `Enter`.

   Espruino will return `2`. Looking good so far.

2. Type `1 / 2`, then press `Enter`.

Espruino will return `0.5`, exactly as we expect. However, many languages don't do this.

JavaScript treats all numbers as if they are fractional (we call this *floating point*), rather than whole numbers, or integers. This means that if you perform an operation that would produce a floating-point (i.e., fractional) value, that's what you'll get. So when we divided 1 by 2, JavaScript returned the floating point `0.5`.

In languages like Python, however, `1/2` will return the answer `0`. This is because `0` is the correct answer if you do the math in whole numbers (and you ignore the fractional parts rather than rounding them). In Python if you want to get a floating-point result you have to make it clear that that is what you want. In this case you'd do that by typing `1 / 2.0`. This isn't needed in JavaScript.

Math with Strings

Math symbols can also control how different strings relate to each other. For example, type `"Hello" + " World"`, then press `Enter`.

Espruino will display `"Hello World"`. The interpreter has identified two strings, recognized that they contain text, and concatenated them.

Math with Strings and Numbers

Next, try typing `1 + "2"`, then press `Enter`.

This time Espruino will return `"12"`, not `3`.

This is because Espruino has recognized the `"2"` as a string. The `"2"` has essentially become a textual representation of a number rather than a number itself. Espruino won't convert the `"2"` back to a number because in many cases it is not possible for a string to be converted into a number (a string could just be text). So instead Espruino converts the `1` into a string, and adds it to the `"2"` by displaying the two strings next to each other, as with `"Hello World"`.

However, something a bit different happens if you type `1 - "2"`, then press `Enter`. This returns `-1`.

In this case, JavaScript knows that it can't subtract one string from another, so it tries to convert `"2"` to a number instead, and succeeds. Having converted `"2"` to a number, JavaScript can then subtract it from `1` and return `-1`.

But what happens if you try to subtract a string that can't be converted into a number? Try typing `1 - "Oops"`, then press `Enter`.

JavaScript returns `NaN`.

What's that? Well `NaN` stands for `Not a Number`. JavaScript failed to convert `"Oops"` to a number, so instead it returned `NaN`. Despite its name, JavaScript treats `NaN` like a number.

This means you can do math with it, although the result will almost always be another `NaN`.

The Helpful Language

So, JavaScript sometimes reads a number as a string, and sometimes reads a string as a number, and it also lets you do math with `NaN`. These are all examples of one of JavaScript's most distinctive features: the way it tries to be helpful. If your code is unclear or ambiguous, then JavaScript will take a best guess at what you mean and proceed on that basis, rather than alerting you to a problem.

This is one of the reasons why JavaScript has been so successful. You probably visit many web pages a day, almost all of which use JavaScript. Many of these pages will have bugs in the underlying JavaScript code, but the JavaScript will ensure that the the code continues to work as well as possible. As the end user, you won't usually be aware that the bugs are there at all.

Variables

A variable is a way to store values inside your program, in a named form so that you can reference them later.

To define a variable in JavaScript, enter `var`, then the name of the variable (without spaces in it). For example, for a variable called `a`, you would have `var a`. This allows you to reference your variable by name when you need it, just by typing `a`.

If you want to assign a value to your variable you need to add `=` followed by your instructions for determining the value of the variable.

ES6

In ES6, there are other keywords called `let` and `const` that are used in a similar way to variables, but have slightly different effects.

1. For example, to instruct a variable to have the value `4`, type **var a = 4**, then press `Enter`.

2. Now if you type `a`, then press `Enter`, Espruino will recall the value of `a` and will display `4`.

3. Now try typing `A`, then press `Enter`.

Espruino will display `Uncaught ReferenceError: "A" is not defined`. This is because the variable requested (`A`) could not be found so an exception (an error) was created. Your code will stop executing at the moment when the exception is created (or "thrown," in coding jargon) and Espruino will display the command prompt and report the exception.

So why did this happen? Well, as you may have guessed, JavaScript is case sensitive. When you define a variable, function, or anything else, the capital and lowercase letters have to match up each time. We have defined the variable `a` as `4`, but have not yet defined the variable `A`. `a` and `A` can be two different variables with two totally different values!

In JavaScript you can't restrict a variable to only ever be one kind of data. So, if you instruct your variable to have the value `4` (a number), you can subsequently change that variable's value to a string, an array, or even a function (we'll come to these shortly).

It is also possible to define a variable without using `var`, for example, by typing `a = 4`. However, this will define a global variable, rather than a local variable. A global variable will apply across your whole program, whereas a local variable will only apply within a particular function. This chapter talks about functions later, but it is worth being aware that if you try to create a variable in a function without using `var` this can lead to a lot of bugs. This is because there is a risk that this global variable will overwrite a variable used elsewhere in the program. Always try to use `var` where possible.

ES6

We mentioned strings before, and ES6 has another way of defining strings called *template literals*. These start and end with the special backtick character, `` ` ``.

If you use the sequence `${an_expression}` inside a templated literal, the result of executing that expression will be inserted into the string. Here are a couple of examples:

```
>`One plus two is ${1+2}`
="One plus two is 3"
```

```
>a=5
=5
>`We set a to the number ${a}`
="We set a to the number 5"
```

In case you were wondering what "expression" means here, this is a coding term that broadly means an instruction for the interpreter to execute. An expression will usually return a value (for example, the expression `1+2` returns the value `3`).

Increment and Decrement

One way of making a variable bigger is to type `a = a + 1`. This simply adds 1 to your original variable. Similarly, to make a variable smaller you could type `a = a - 1`.

However, this means we have to write out our variable twice, which can be problematic if we have a really long variable name: `our_really_long_variable = our_really_long_variable + 1`. This duplication can lead to bugs and slow execution.

Instead, JavaScript (like many other languages) has "increment operators," which appear as either `+=` or `-=`. These allow you increase or decrease your variable by typing your variable, then the appropriate increment operator, then a number. For example:

- `a = a + 1` is the same as `a += 1`.

- `a = a - 1` is the same as `a -= 1`.

This also works for other operators such as `*=`, `/=`, `|=` (which we'll come to later), and others.

Adding and subtracting just 1 are very common operations, so there are other operators that help you do this in other ways.

`a++` adds one to the variable `a`, returning the old value of `a` (it works with `a--` too). This is often used in `for` loops, which we'll come to later.

```
>var a=1
=1
>a++
=1
>a++
=2
>a
=3
```

Just to make things difficult, `++a` also adds one to `a` but, unlike `a++`, `++a` returns the new value of `a`. This also works with `--a`.

```
>var a=1
=1
>++a
=2
>++a
=3
>a
=3
```

 When you're not using the result of an increment you can use `a++` or `++a` interchangeably, since the only difference is in the value they return, not the operation they perform.

Objects

In JavaScript, objects are containers that can hold a variety of named variables or "properties."

1. For example, type `a = { one : 1, two : 2 }`.

   In this example, we have used the semicolons to name the property `1` "one", and the property `2` "two."

   You can then access these items using the dot (`.`) operator.

2. Try typing `a.one` to read the item named `one`.

This returns `1`.

3. Now we'll try adding a new property into the object. Type `a.three = 3`.

    Note that if you try to access a property like `a.four`, which we haven't defined yet, you won't get an error. Instead, Espruino will just return `undefined`.

4. You can also refer to your property indirectly using square brackets. Try typing `a["one"]`. This will also return the name of your property, `1`.

5. Or to make things even more complicated, try typing `var b = "one"` and then `a[b]`. This will return `1`. We have given the variable `b` the value `one`, so `a[b]` is now the same as typing `a["one"]`.

Functions

Now that we've looked at simple math, let's call some functions. Functions are named bits of code that do a particular job and have usually been written beforehand.

parseInt

When doing some simple math, we learned about the difference between a number (or integer) like 42 and the string "42". The function called `parseInt` can be used to convert strings into integers. To give, or "pass" data (which we call "arguments") to a function, put the arguments into parentheses after the function like this: `parseInt("10")`.

This passes the string `"10"` to the function `parseInt`. The function then transforms the string `"10"` into an integer and returns `10`. This means that you can now work with the string (`"10"`) as though it was an integer (`10`). For example, you can now do math with the string. Previously, if you had typed `"10" + 10` this would have returned `"1010"`, but now you can type `parseInt("10") + 10`, and this will return `20`.

Functions in Objects (Methods)

We looked at some simple variables earlier, but it is important to be aware that variables can be much more complex. For example, variables can be objects, functions, or arrays. Some global variables contain a variety of useful properties and functions.

For instance, JavaScript has an object called `Math`. This object contains a number of arithmetic properties and functions.

Just as with the object we defined earlier, the (`.`) operator allows you to access a property within it; for example, typing `Math.PI` displays `=3.141592...`, because the value of pi, or π, is 3.1415….

Because functions can also be properties of objects, you can also call functions using the (`.`) operator.

Strictly speaking, a function that appears within an object is called a method.

Typing `Math.abs(-2)` displays `2` because the function `abs` returns the absolute value of the argument—the number's distance from zero, in either direction. Using absolute value has the effect of turning all negative numbers into positive ones.

console.log

One of the most useful functions is `console.log`. This function allows you to display text in the console area while your program is running, and is very useful for debugging. For example, `console.log` can be used to display the value of a variable at any given time during execution, which allows you to check whether your program is running as you expect it to.

```
>var a = 42;
=42
>console.log("The value of a is ",a)
The value of a is   42
=undefined
```

 While not part of standard JavaScript, Espruino provides the `print` function. This behaves identically to `console.log` but is just a little faster to type!

Defining Functions

Now you know how to call functions, you can make your own:

1. Type `function add(a,b) { return a + b; }`.

   This creates a function called `add`, which needs to be passed two arguments `a` and `b`. When the function is called, the code inside the curly braces (`{ }`) will be executed. In this case the code says that we want to `return` the result of adding `a` and `b` together.

2. So now try calling the `add` function using the arguments `1` and `2` by typing `add(1,2)`.

   This returns the value `3`.

Functions can be written with code on multiple lines. As with many other programming languages, if you do this you should separate statements with semicolons (`;`), even if it isn't always mandatory in JavaScript.

A "statement" is just a piece of code. Usually, the word "statement" is used to describe a piece of code that does something, but doesn't return a value (for example, the statement $a=1$ defines a as 1).

In JavaScript it is possible to define one function inside another. The inner function will be able to access variables defined in the outer function. An inner function that does this is called a closure.

For example, if you define the following function:

```
function outer() {
    var hello = "Hello World!";
    function inner() {
        return hello;
    }
    return inner;
}
```

Then type:

```
var inner = outer();
inner();
```

This returns `Hello World`.

When we call the function `inner` this refers to the inner function `function inner() { return hello; }`, however this in turn refers to the outer function, which has defined `hello` as `"Hello World!"`.

This is a bit confusing, but using closures can be very useful as it helps to save memory. When the `inner` function has been executed, all the variables belonging to it (like `hello` and `"Hello World!"`) will be automatically removed from memory.

Inline Functions

In some cases you might want to define a function to be used just once, in which case we can define it inline, and there's no need to give it a name.

For example, here we're calling `setTimeout` to schedule a function to be called two seconds in the future. We're not calling the function from anywhere else, so there is no need to give it a name:

```
setTimeout(function() {
    console.log("Hello");
}, 2000);
```

Events and JavaScript

JavaScript itself uses functions heavily because it is event based. Rather than having your code run in one big loop, in JavaScript you write functions that you want to be executed when an event occurs.

If you're writing a web page and you want something to happen when the user clicks a button, you define a function and then configure the button so that your function is called when it is pressed. The same is true when you're using Espruino with real, physical buttons.

Working with events has some huge benefits. The first benefit is power consumption: if the JavaScript interpreter has control when your software is idle, it can put your microcontroller to sleep and save power.

Another benefit is the ability to multitask. In normal, procedural Arduino code you might write a function like this to flash an LED on and off:

```
void flashLed(int ledPin) {
  digitalWrite(ledPin, 1);
  delay(1000);
  digitalWrite(ledPin, 0);
  delay(1000);
}

flashLed(LED1);
```

However, in Espruino, you could write:

```
function flashLed(ledPin) {
  digitalWrite(ledPin, 1);
  setTimeout(function() {
    digitalWrite(ledPin, 0);
  }, 1000);
}

flashLed(LED1);
```

Here, we're turning the LED on and then defining an inline function that will turn the LED off. We pass that to a built-in function called `setTimeout`, which schedules the inline function to be called in 1 second, which then turns the LED off.

So what is different? The `flashLed` function in the Arduino code takes a whole 2 seconds to execute, during which time it's very difficult to execute any other code.

However, the Espruino code executes two functions very quickly but 1 second apart, and is available to execute other functions for the rest of the time.

If we now wanted to flash two LEDs at the same time (not one after the other), in Espruino we could simply write:

```
flashLed(LED1);
flashLed(LED2);
```

However, in Arduino you would have to rewrite the `flashLed` function. If you wanted to flash one LED and then half a second later start flashing the other, you could easily do that as well with Espruino, also without having to change the `flashLed` function:

```
flashLed(LED1);
setTimeout(function() {
  flashLed(LED2);
}, 500);
```

This can be amazingly powerful, and is made possible by the use of closures and inline functions as just described.

Arrays

Arrays are very similar to objects, except they have the concept of a length. You can create an array very easily:

1. Type `a = [5,8,3,7]`.

   This creates an array with four items (you can just use `[]` to create an empty array).

   You can now access an element in the array (they're numbered from `0`).

2. Type `a[1]` and you'll get `8`; `a[3]` will return `7`. If you type `a[50]` (thereby accessing an item that doesn't exist), you'll get `undefined` returned rather than an error.

You can add items onto the end of the array in two ways:

1. You can just assign an element to the end; type `a[4] = 40`.

2. Or you can use the `push` method; type `a.push(45)`.

3. You can now check the length of the array with `a.length`.

   `6` will be reported, because we just added two more elements.

4. To remove an item from the end of the array, type `a.pop()`.

   `45` will be returned, and the length will drop to `5`.

5. To remove an item from the beginning of the array (shifting everything else along), you can just type `a.shift()` and `5` will be returned.

   To put something back onto the beginning, just use `a.unshift("Another ele ment")`.

6. Arrays also have many useful functions for iterating over them. `Array.forEach` will call the function in its argument for each element in the array. Try typing the following:

   ```
   var a = [5,8,3,7];
   var sum = 0;
   a.forEach(function(x) { sum += x });
   console.log(sum);
   ```

   It will print `23`, the sum of all elements in `a`.

7. You can use `Array.map` to return a new array made from the return values of the supplied function when it is called for each array element. Type `a.map(func tion(x) { return "Hello "+x; });`.

`[ "Hello 5", "Hello 8", "Hello 3", "Hello 7" ]` will be printed. This can be a really useful tool.

Sometimes you want just one result rather than an array, and in this case you can use `Array.reduce`.

8. Type `a.reduce(function(accum,value) { return accum + value; }, 0);`.

Espruino will return `23` (as the previous example did, but just with a little less code).

So what's happening? Espruino is calling your function repeatedly, passing the result of the last call into the next. It's a bit like the following code, but Espruino is doing everything automatically:

```
var a = [5,8,3,7];
var sum = 0;
function summer(accum,value) { return accum + value; }
sum = summer(sum, a[0]);
sum = summer(sum, a[1]);
sum = summer(sum, a[2]);
sum = summer(sum, a[3]);
console.log(sum);
```

ES6

In ES6, there is a shorthand way to define inline functions called an *arrow function*. This code:

```
setTimeout(function() {
  console.log("Hello");
}, 2000);
```

could be rewritten like this:

```
setTimeout(() => console.log("Hello"),
2000);
```

It's really handy to use arrow functions with `Array.map` and similar—for example, the follow-ing code will return a new array with `1` added to every element:

```
[1,2,3,4,5,6,7].map( a => 1+a );
```

Or this will sum all the elements in an array:

```
[1,2,3,4,5,6,7].reduce( (b,c)=>b+c, 0 )
```

Look out for the arrow (`=>`) in modern ES6 code to get an idea where functions are defined.

Object Orientation

One of the main places JavaScript differs from other languages is that functions are themselves objects, and can have properties added or removed.

JavaScript defines a special property called `prototype`. If this is defined, all objects of that *type* will search that type's `prototype` property for properties that don't exist in the main object.

As an example, `"Hello"` is a `String`, so we can add a function to `String.prototype`. Within this function, the special variable `this` represents the original object.

1. Enter the following to add a function called `print` that will print a `String` object to the console:

```
String.prototype.print = function() {
    console.log(this);
};
```

2. Type `"Hello".print()` and the `String.prototype.print` function will now be called, and will output `Hello`.

   `.print()` can now be called on any `String`.

3. We can make a new type of object ourselves. To start, create a function:

```
function Person(name, age) {
    this.name = name;
    this.age = age;
}
```

4. `Person` will be our type of object. We can now create `Person` objects with:

```
var alice = new Person("Alice", 42);
var bob = new Person("Robert", 37);
```

5. You can access properties directly, for example with `bob.age`. If we want to share some code between them we can add a function to the `prototype`, as we did with `String` earlier:

```
Person.prototype.introduce = function() {
    console.log("This is "+this.name+", they're "+this.age+" years old");
};
```

And now, you can run `alice.introduce()` and `This is Alice, they're 42 years old` will be displayed.

Bitwise Arithmetic

Bitwise math allows you to easily fiddle with the bits that make up numbers—the basic building blocks of computers. Many developers will never have a need to use bits, but because we're dealing with hardware you'll probably want to be at least slightly familiar with them!

In JavaScript, numbers are usually fractional (as we covered earlier), but the bitwise arithmetic operations convert numbers into 32-bit integers internally, before doing anything with them.

First, let's see which bits are in a number:

1. Type `(100).toString(2)`.

This will convert the decimal number 100 to a string representing a base 2 number (binary): `"1100100"`.

Why Did We Put 100 in Parentheses?

If you type `100.` in JavaScript, the interpreter will assume that you meant to enter a fractional number. Instead, you need to put the parentheses around it so that when `.` is encountered, JavaScript knows it's no longer expecting fractional digits, but a property of the number.

You could also write `100 .toString(2)`, but writing code where the number of blank spaces changes what it does is generally considered to be a bad idea.

So what does `1100100` mean? In decimal, the position of each digit represents some power of 10, so we have 1000s, 100s, 10s, 1s, etc. In binary, the position of each digit is a power of 2. We have (working from the left to the right), 64s, 32s, 16s, 8s, 4s, 2s, 1s. For the binary number `1100100` we have $1*64 + 1*32 + 0*16 + 0*8 + 1*4 + 0*2 + 0*1 = 100$.

There are, however, much better ways of writing binary numbers down. The first way is to use `parseInt` as we did previously. If given a seconds argument (the base of a number), `parseInt` will convert a text string in that base into an integer.

1. Type `parseInt("1100100",2)` and `100` will be displayed.

2. But in more modern versions of JavaScript, you can write a binary number and just prefix it with the text `0b`. Type `0b1100100` and `100` will be displayed again.

ES6

Though we're covering binary numbers like `0b1100100` here because they're really useful, binary (`0b`) and octal (`0o`) are actually part of the ES6 spec, so while they will work on Espruino they're not guaranteed to work on every web browser.

Hexadecimal

But why stop with binary and decimal? Why not use some other number systems too?

Hexadecimal (base 16) is very popular in computer code, because it lets you get 4 bits of data into one single digit, and it's easier to see what a specific bit is set to by looking at it. To do that the digits `0` to `9` are used as you'd expect, but then the letters `a` to `f` are used for `10` through to `15`.

To write a hexadecimal number, you prefix it with `0x` (because he**X**adecimal). For example `0xa` is 10, and `0x10` is 16.

3. This brings us to an annoying gotcha in JavaScript. Type **0999** and press `Enter`.

   As expected, it writes `999`.

4. Type **0123** and press `Enter`.

   Espruino will write `83`—but why? For historical reasons, numbers starting with `0` are treated as octal (base 8).

Since `999` couldn't possibly be octal, JavaScript "figures it out" for you, and silently treats the number as decimal. But `123` is valid octal, so JavaScript decodes it as `1*64 + 2*8 + 3`.

As a result, you should never put a `0` at the front of your nonfractional numbers in JavaScript (or any programming language), as it's amazingly rare that you ever actually intend to define an octal number (and if you do you could do so in ES6 with `0o123`).

Bitwise Operators

So now that we know how to define binary numbers, we can look at doing some bitwise math. It's called bitwise because the operations generally act on each bit in a number individually (an `add` operation has to *carry* to the bit on the left if a digit overflows).

The *binary and* operator, written as `&` (and not to be confused with `&&`) does what you'd expect from the name. Each bit in the output is equal to `1` only if the corresponding bit in the first argument is `1` *and* the bit in the second is `1`. If either one of the two bits is a `0`, then the output bit will be `0`.

1. Type `(0b1010101010101 & 0b0001111110000).toString(2)`.

   `"101010000"` is returned, because the second argument has effectively *masked* off the first. So why weren't more zeros output on the left of the result? Much as when you write a decimal number, you don't generally put any zeros at the front of it if you don't have to:

   A | 1 | 0 | 1 | 0 | 1 | 0 | 1 | 0 | 1 | 0 | 1 | 0 | 1

B	0	0	0	1	1	1	1	1	1	0	0	0	0
A and B	0	0	0	0	1	0	1	0	1	0	0	0	0

2. Now, we can try *binary or*, written with a pipe symbol (`|`) (and again, not to be confused with `||`!).

   This also does what you'd expect based on the name. Each bit in the output is equal to `1` only if the corresponding bit in the first argument is `1` *or* the bit in the second is `1`. If both bits are `0`, the result will be `0`.

3. Let's use the same numbers and replace the operator. Type `(0b1010101010101 | 0b00001111110000).toString(2)`.

   Now, you get `"1011111110101"` because, as expected, everywhere there's a 1 in the input (no matter which argument it was in), there's a 1 in the output:

A	1	0	1	0	1	0	1	0	1	0	1	0	1
B	0	0	0	1	1	1	1	1	1	0	0	0	0
A or B	1	0	1	1	1	1	1	1	1	0	1	0	1

4. Then there's *exclusive or*, or *xor*—written as `^`. It kind of makes sense when you think about it, but it's still a bit of a confusing name. An output bit is `1` only if *one* of the output bits is `1`, but it's so exclusive that if both are `1`, the output will actually be `0`. Type `(0b1010101010101 ^ 0b00001111110000).toString(2)`.

   You get `"1011010100101"`, so what happened? Well, where the righthand argument's bits were `0`, the lefthand argument's bits stayed the same, but where they were `1`, the bits on the lefthand side got flipped. These operators don't give one argument priority over another, so you could just as easily think of it as working the other way around too!

A	1	0	1	0	1	0	1	0	1	0	1	0	1
B	0	0	0	1	1	1	1	1	1	0	0	0	0
A xor B	1	0	1	1	0	1	0	1	0	0	1	0	1

5. It's easier to see what happens with different numbers:

```
>(0b1111111111111 ^ 0b00001111110000).toString(2)
="1110000001111"
```

```
>(0b0000000000000 ^ 0b0001111110000).toString(2)
="0001111110000"
```

6. Finally, there's one simple operator, the *not* operator, written as a tilde (`~`). This takes just one argument and flips every single bit. You could think of it as an *xor* with a 32-bit number full of `1`s.

Type `(~0b0001111110000).toString(2)`.

You get `"-1111110001"`. So what happened here? For signed numbers (numbers that can be negative), when the biggest (leftmost) bit of a number gets set (in this case the 32nd) then the number is treated as negative. It's as if instead of being responsible for adding `2^31`, it subtracts `2^31` instead. And in this case, JavaScript sees that the top bit is now `1` and decides to treat the number as being negative rather than just printing the value of all 32 bits as we wanted.

Why Is There a 1 on the Right?

Inverting all the bits of a number has the same effect as negating it and then subtracting one.

Try it out with 4-bit numbers as it's easier to add up! The leftmost bit will represent `-8`, the second left `4`, then `2`, then `1`.

0000→0 1000→-8 0001→1 1001→-7 0010→2
1010→-6 0000→3 1000→-5 0000→4 1000→-4
0000→5 1000→-3 0000→6 1000→-2 0000→7
1000→-1

So if you take `1100` (-4) and flip all the bits you get `0011`, which is 3.

In order to put the minus sign in what was printed ("-1111110001"), JavaScript negated the number, which ended up negating all the bits and subtracting one.

To get something readable, let's use `&` to mask off that pesky top bit. We could write a binary number with `0` followed by 31 `1`s, but that would be really painful.

Instead, we'll use hexadecimal (see the earlier discussion). In hexadecimal the number is just `0x7FFFFFFF`; it's much easier to see what's happening than if we wrote `2147483647` (the decimal equivalent) while it is more compact than the binary, which is `0b01111111111111111111111111111111`!

Type `(0x7FFFFFFF & (~0b0001111110000)).toString(2)`.

And now, we get `"1111111111111111111110000001111"`, showing 31 of our 32 bits as they actually are, but without the negative sign, which was caused by the top bit.

Bit Shifting

Now, we get onto bit shifting. These are also operations that don't get used a lot in everyday programming unless you're dealing with hardware!

Sometimes you may want to move all the bits of a number left or right. Hardware inside (or connected to) a microcontroller is often trying to be very efficient, and in order to save memory (or complexity in the hardware) one area of memory may be used to represent multiple different things.

For instance, in many LCD displays a single 16-bit number (representing a value between 0 and 65535) may be used to represent a color. Five of the bits are used for red, six for green (because the eye is more sensitive to green), and five for blue. To construct the 16-bit number you'll need to use bit shifting to shift the bits for each of red, green, and blue into the right places for that 16-bit number.

In decimal, you can shift digits around by multiplying and dividing. `1234` can be shifted left by one digit by multiplying it by `10`, making `12340`. Dividing by `10` will shift it back to the right. The same is true in binary, but because the number system is base 2, you multiply or divide by `2`.

In a computer it is very fast to shift bits around as numbers are already represented in binary; however, multiplication and especially division take a lot more time. Because of this there are special operators you can use to perform shifting directly.

The main operators are *shift left* (represented by `<<`) and *shift right* (`>>`). The first argument is the number to shift, and the second is the amount of bits.

Shifting by 0 does nothing…

```
>(0b11110000 << 0).toString(2)
="11110000"
```

Shifting left moves bits left (filling the gap with 0), which makes the number bigger. For every one place you shift, the number gets twice as big:

```
>(0b11110000 << 1).toString(2)
="111100000"
>(0b11110000 << 4).toString(2)
="111100000000"
```

And shifting right makes the number smaller. Any bits that are shifted off of the right of the number will disappear, so if you shift so far right that all the bits fall off the end, you get 0:

```
>(0b11110000 >> 3).toString(2)
="11110"
>(0b11110000 >> 4).toString(2)
="1111"
>(0b11110000 >> 8).toString(2)
="0"
```

Or do you? It would be nice if when you shifted a negative number right, it got half as big as well. Remember what we said earlier about the top bit of the number making it negative? If you were to blindly shift all bits in a negative number right, the `1` in the top bit would move to the right and the number would stop being negative, and would instead become a very big positive number.

To get around this, the normal shift right retains the top bit's value when it shifts right, so `-0b11110000 >> 4` really will be `-0b1111` as you'd expect.

Sometimes you want to treat the number as if it can't be negative though, and you just want to shift `0` bits in (like you do when shifting left). To get around this, JavaScript has `>>>`, which is called *unsigned shift right*.

Unsigned means the number doesn't use the leftmost bit to store the sign (whether it is negative or not), so the number can only ever be greater than or equal to 0.

For positive numbers it's exactly the same:

```
>(0b11110000 >>> 3).toString(2)
="11110"
>(0b11110000 >>> 4).toString(2)
="1111"
>(0b11110000 >>> 8).toString(2)
="0"
```

But for negative numbers it's different:

```
>(-0b11110000 >>> 3).toString(2)
="11111111111111111111111100010"
>(-0b11110000 >>> 4).toString(2)
="1111111111111111111111110001"
>(-0b11110000 >>> 8).toString(2)
="111111111111111111111111"
```

Do Other Languages Have `>>>`?

Often other languages don't have to have a special shift operator. For instance in C, you have to define the exact type of your numbers and variables: how many bits they have, as well as whether they can be negative (signed) or are always positive (unsigned). The compiler then knows if a number is supposed to be signed or not and it can use the correct operation wherever `>>` is used.

If Statements

`if` statements in JavaScript work the same way they do in other languages. If the expression in parentheses evaluates to `true`, the code after it will be executed; otherwise, if there's an `else` after it, that gets executed if the expression is `false`.

1. `if (true) console.log("This is running")` will print `This is running`.

2. `if (false) console.log("This isn't running")` won't print anything.

3. But the following will:

```
    if (false) {
      console.log("This isn't running")
    } else {
      console.log("This is running")
    }
```

Instead of using the values `true` and `false`, you can use variables directly. For instance, the number `0` is treated as `false`, and nonzero numbers are treated as `true` (`NaN` and `undefined` are also treated as `false`).

Confusingly, an empty array (`[]`) or object (`{}`) is treated as `true`, while an empty string is treated as `false`:

```
    var name = "";
    if (name) {
      console.log("I have a name! It's "+name);
    }
```

Often you'll want to be able to check a variable is equal to something; for instance, we can use the equals operator `==` we covered before:

```
    if (a==42) {
      // Do something
    }
```

This also leads us to another tricky problem: typing `"42" == 42` returns `true`. The two arguments aren't identical though: one is a string and one is an integer. They just happen to represent the same value.

If you care about this you can use a special `===` operator. Three equals shows you really mean it, and `===` checks the value of the variable as well as the type, so `"42" === 42` will return `false`.

If you want to do the opposite you can use the *not* (`!`) operator at the start of an expression. This will negate a boolean value, turning `true` into `false` and vice versa. For instance, `!(a==3)` will be true if `a` isn't 3, although in this case you could write `a != 3` more succinctly (`a !== 3` works for checking the type as well).

&& and ||

In many cases you would want to do something if something is true *and* something else is true. You can do this with the *and* (`&&`) and *or* (`||`) operators. While these are read out the same as the bitwise operators covered earlier, they behave slightly differently.

For starters, `&&` and `||` work on boolean values, not on each bit. Their behavior is also subtly different to most other programming languages:

- For most languages, `a && b` (or its equivalent) follows the rule: "If a is true and b is true, return true, else return false."

- In JavaScript, `a && b` means: "If a is true and b is true, return b, else return a."

  Similarly, `a || b` means: "If a is true then return a, else return b."

While this behavior turns out to be identical to other languages when you're dealing with `if` statements, it means that these operators have other uses as well.

For instance, when looking at JavaScript code on the internet, you might see something like: `a = a || "default_value"`.

This means "if a isn't false then use a, else use the default value," which is a tidy way of setting defaults where nothing has been specified (although you should be wary of what happens if `a` is defined as `0`!).

Ternary Operators

Ternary operators (`a ? b : c`) are very powerful, but can often be hard to understand when you see them written in code.

Very often, you'll want to use one value in one case, and one value in another.

You might write:

```
if (a)
   value = 1;
else
   value = 2;
call_a_function(value);
```

But using a ternary operator you can just write `call_a_function( a ? 1 : 2 );`.

The ternary operator looks at the expression before `?`, and if it's `true`, it returns the expression before `:`, but if it's false it returns the expression after it.

Going back to the problem with `a = a || "default_value"`, when `a` is set, but is `0` or `false`, we can now use ternary operators to be much more explicit about when `a` should be used, or when a default value should be used:

```
a = (a!==undefined) ? a : "default_value";
```

for Loops

`for` loops are an easy way to loop over things multiple times, and in their basic form behave just like in C, Java, PHP, and other languages.

A basic `for` loop is of the form:

```
for (initialise ; compare; iterate) {
   do_some_work
}
```

and might look like this:

```
for (var i = 0; i<2; i++) {
  do_some_work
}
```

This will do the following:

1. Set the variable `i` to `0`.

2. Check if `i < 2`. It is, so carry on.

3. Run `do_some_work`.

4. Increment `i`. It's now `1`.

5. Check if `i < 2`. It is, so carry on.

6. Run `do_some_work`.

7. Increment `i`. It's now `2`.

8. Check if `i < 2`. It isn't, so leave.

There is a second form of `for` loop in JavaScript, and it looks like `for (var i in array) ...`.

This calls the code after with every element's name in the array. For example, this:

```
var arr = ["ham", "eggs", "cheese"]
for (var i in arr) {
  console.log(i, arr[i]);
}
```

will output:

```
0 ham
1 eggs
2 cheese
```

Note that `i` isn't set to the array's contents, but to the index of each item in the array.

There are also `do` and `while` loops in JavaScript, but these aren't used in this book so we won't cover them.

There are other ways to iterate if you have an array. For instance, we covered the `forEach` method earlier:

```
var arr = ["ham", "eggs", "cheese"]
arr.forEach(function(value, key) {
  console.log(key, value);
});
```

Exceptions

At some point during these exercises, you might have typed something wrong and got a message like this:

```
Uncaught SyntaxError: Got '#' expected EOF
```

This is an exception. The JavaScript interpreter has found something that it doesn't like, and stops execution where the problem occurred.

You can force it to stop execution by *throwing* an exception:

1. Type `throw new Error("Oh No!")`.

   You'll see `Uncaught Error: Oh No!`

2. By themselves, exceptions are not very useful, but you can `catch` exceptions, so if your code breaks then you can do something about it by typing the following:

   ```
   a = '{"Hello":"World"}';
   try {
     console.log(JSON.parse(a));
   } catch (e) {
     console.log("Something went wrong:"+e)
   }
   ```

   In this case, everything will work fine, as `JSON.parse` could execute the JSON-formatted string `a`.

3. Type `a = '{Hello:"World"}';` (which is badly formatted JSON), press ↑ to step back in history, and call the `try` statement again.

   Now `Something went wrong:SyntaxError: Got ID:Hello expected '}'` will be displayed. You've been able to carry on executing code even though `JSON.parse(a)` threw an exception because of bad data.

And that wraps up the introduction to JavaScript's main features: you're ready to make some stuff!

PART II

Motors

If we want to move things in the real world from our microcontroller, we're going to need a motor.

In these chapters, we'll learn how different kinds of motors work, and how to make some ourselves.

What Is an Electric Motor?

An electric motor is a device that turns electrical energy into mechanical energy. In 1821, Michael Faraday demonstrated what was perhaps the first example of an electromagnetic motor (see Figure 5-1). A wire was dangled into a pool of mercury and placed next to a permanent magnet. When a voltage was applied between the mercury and the top of the wire, current flowed. It created a magnetic field, and the end of the wire started orbiting the magnet.

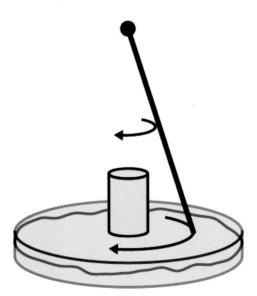

Figure 5-1 *Faraday's electric motor*

Do not try to build this motor. Mercury is extremely toxic, and even exposure to the vapor can cause long-lasting damage to the lungs, brain, and kidneys.

*As late as the 20th century, mercury was still handled without adequate safety precautions, and caused severe health problems for many involved. The use of mercury in the treatment of animal fur for hats led to **Mad Hatter Disease**, the basis of the **Mad Hatter** character in Lewis Carol's **Alice in Wonderland**!*

The Earliest Motor

The *first* electric motor was actually created by Andrew Gordon (a Scottish monk) in the 1740s. Unlike most motors today, which are electromagnetic, Gordon's motor was electrostatic (see Figure 5-2). It used the attraction caused by high voltages (much like a balloon will stick to the wall once charged up against a sweater) to turn a rotor.

With the invention of chemical batteries, electromagnetic motors replaced electrostatic motors. However, electrostatic motors are still extremely useful at small scales, and the advent of tiny microelectromechanical systems (MEMS) such as the accelerometers and gyros in your mobile phones means that electrostatic motors have increased in popularity.

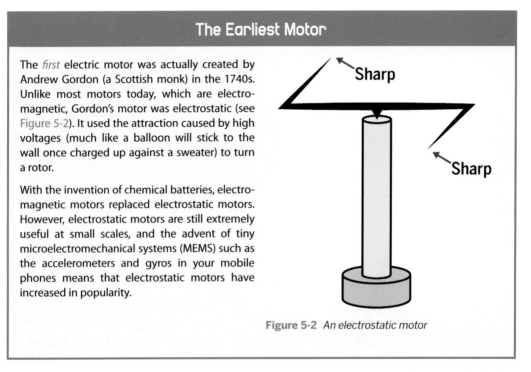

Figure 5-2 *An electrostatic motor*

Obviously, re-creating Faraday's mercury-based experiment as it was done is a bad idea without a lot of ventilation. However, with modern rare-earth magnets and a bit of lateral thinking, you can reproduce Faraday's motor easily at home without risking poisoning!

Experiment 1: Faraday's Motor

You'll need the following (see Figure 5-3):

- Neodynium disc magnets (with as large a diameter as possible)
- An AA battery
- A 25cm length of single-core wire without insulation

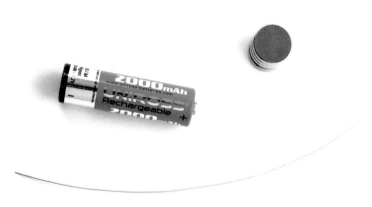

Figure 5-3 *The parts you'll need to make a homopolar motor*

To re-create Faraday's motor:

1. Place the disc magnet onto a nonmetallic tabletop. Stacking multiple magnets on top of each other (as I have done in Figure 5-3) will make your motor a bit easier to assemble.

2. Place the AA battery with the positive end pointing downward, on top of the magnet.

 Some AA batteries don't have a positive end that's attracted to magnets. You'll need to find one that does!

3. Now, take the length of wire and bend it in two.

4. Fold the two sides out and into a tall M-shape:

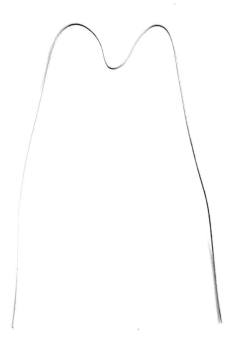

5. Fold the ends inward so that they overlap by around 1cm, making sure that the height of the *M* matches the height of the AA battery plus magnet.

6. Now carefully place the tip of the *M* in the wire over the negative terminal of the AA battery, making sure the two ends of the wire rest gently against the magnets:

You may need to fiddle a little to make sure you get a good electrical contact while not applying too much force with the wire, but you should notice the wire wants to start spinning.

I found that my motor spun so quickly the wire fell off the top of the battery, so I added a small washer to help keep it in place:

So how does this work? It uses the Lorentz force. Current flows around the circuit that's been made: through both sides of the wire, the battery, and the magnet. Fleming's right-hand rule says that if you extend the middle finger, index finger, and thumb on your right hand at right angles to each other (Figure 5-4), each of the three fingers can represent the direction of current, magnetism, and force (it doesn't matter which is which).

Figure 5-4 *Fleming's righthand rule*

The current in the vertical section of the wire runs almost parallel to the magnetic field, so this means that it isn't generating much force. However, current also flows radially outward from the center of the magnet, and this interacts with the vertical magnetic field and causes a force that is perpendicular to both the magnetic field and the current, which pushes the coil of wire around (Figure 5-5).

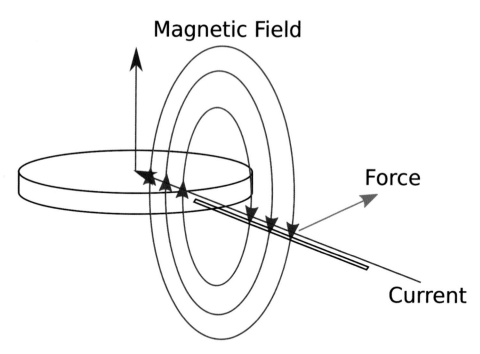

Figure 5-5 *How force is generated from the current flowing in the motor*

The current does flow in the opposite direction at the top of the piece of wire, but the magnetic field is much weaker there, and so produces less force in the opposite direction.

Do modern motors work like this? Not quite. Faraday's motor and the one you just made are *homopolar motors*, which means that the magnetic field does not change. While these motors are very interesting, it has been hard to produce them in a configuration that produces a lot of power.

To make a motor more powerful, it is much easier to use the attraction between magnets. If you have two magnets (one fixed, and the other on an axle), the one on the axle will rotate until its north pole faces the south pole of the other because opposites attract.

However, in order to keep the axle keep rotating, you need to change the magnetic field on one of the magnets so that instead of attracting, the magnets will start to repel each other, continuing the rotation until opposite poles are near to each other again.

You can't do this with a normal magnet, but you can do it very easily with an electromagnet just by changing the polarity of the voltage applied to it. You just need a way to change that polarity (see Figure 5-6).

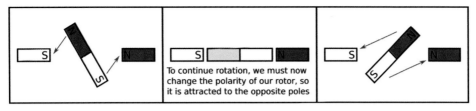

Figure 5-6 *Making a motor rotate by changing the polarity of a magnet*

You can run your motor off of AC power, which changes polarity at a set rate, but this means your motor will only run at a fixed speed (and may be hard to get started). If we have DC power, we're going to need some kind of switch that applies the correct power to the electromagnet at the correct point in the rotation of the motor.

This is called a *commutator*, and it's just a rotating switch (Figure 5-7). Just as a valve in a steam engine lets steam in and out of a piston at certain points in the cycle, conductive *brushes* make contact with metal areas on the commutator, feeding current to the electro-magnet at the correct times in the cycle to make it rotate.

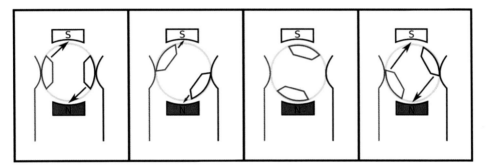

Figure 5-7 *A simple DC motor's commutator*

Experiment 2: Motor with Commutator

You can make a simple electric motor with a commutator yourself. You'll need the following (see Figure 5-8):

- Two neodynium magnets
- Four large bare metal paperclips
- 1.5m insulated copper wire
- A block of wood and four screws
- A cork from a bottle of wine (I'm using a fizzy wine cork because it's bigger)
- Two 2-inch nails
- Sellotape

- A source of DC power: around 6–12v (see Appendix A for ideas)

Figure 5-8 *The parts you'll need to make a simple motor*

Let's create the motor:

1. Push the two 2-inch nails into the ends of the cork:

2. Wrap the wire around the cork lengthways, leaving around 5cm of wire on each end (you may need to cut one end to size):

3. Wrap a bit of sellotape around the nail at the end where the ends of the wire are. This will help to insulate it.

4. Strip both ends of the insulated copper wire, leaving 1 cm of insulation near the cork:

5. Now fold the exposed metal wire in half and twist it together. This will create the contacts of the commutator.

6. Push the twisted bits of metal flat, and wrap a little more sellotape around the very ends of the looped wire to hold the two loops in place.

7. Now cut the excess sellotape off at the end of the nail, leaving a little ring of bare metal. This will help to guide the rotor as it is spinning so it doesn't move to one side or the other:

8. Next, make a stand for each end of the rotor. Fold two paperclips like this:

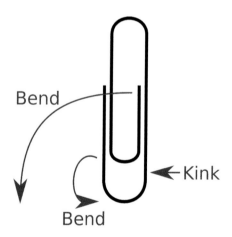

To get this stand:

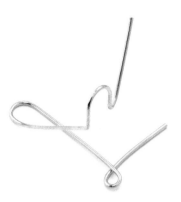

9. Now screw each paperclip into each end of the wooden block, so that you can rest the rotor into the *V* shape in each paperclip.

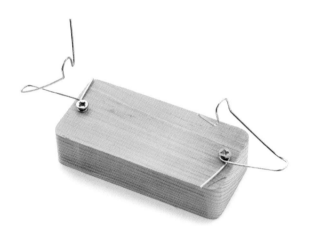

10. Fold the remaining two paperclips so that they look like this:

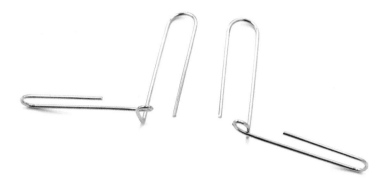

11. Add the rotor and screw the two paperclips down to the block of wood next to each other, so the long length of each paperclip is touching the commutator (you may have to do a little bending to get everything properly aligned!):

12. The final step is to add the magnets. For this motor, the magnets need to be placed either side of the rotor, with the north pole on one side and south pole on the other.

In my case, I added two bits of wood to the side of the stand and glued some magnets salvaged from an old hard disk to them (see Appendix A), but even holding the magnets with your hands will do.

13. Now connect your power source to the two paperclips sticking out of each side of the motor, and spin the rotor by twisting the end of the nail sticking out of it—if all is well, it should start to spin! If not, try bending the two paperclips, making sure they touch the commutator, but with very little force! You may also need to experiment with the positioning of the magnets:

As you might have noticed with your motor, it usually needs pushing to get started. This is partially because of our slightly iffy homebrew commutator, but is also because if the coil we made is perfectly facing the permanent magnet, no rotational force will be applied and so the coil won't be able to rotate at all.

To get around this, most motors use a second electromagnet that is 90 degrees away from the first one, and they have more contacts on the commutator (Figure 5-9). This means that there will always be an electromagnet that isn't facing the permanent magnet, and the motor will always be able to start rotating without human intervention!

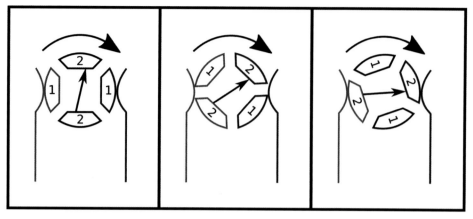

Figure 5-9 *Using a second coil to allow the motor to start from any position*

Of course, more coils can be added to make the motor run more smoothly.

You might have noticed when you ran your motor that there were small sparks coming off the commutator (the same flashes of light you might see if you use a cordless drill and release the trigger quickly). This is because a commutator works by rubbing one conductive material against another, and that can cause sparks, and more serious problems, as the commutators lose contact and regain it. Commutators can wear out if the motor is used for a long period of time, and you can't put as much power through the commutator as you could if the coils were connected directly.

The relatively delicate coils of wire are also spinning around at high speed, when your nice simple permanent magnet is stationary. It seems like it would be much simpler to spin the magnet around while keeping the coils stationary.

You can, but now you can't easily use a commutator to change the voltages on the coils and you're back to square one. What you need is another way to control when and how the electromagnets turn on without rubbing two bits of metal against each other, and this is when—finally—you get a chance to use microcontrollers.

Brushless DC Motors

In brushless DC motors, you have coils that don't move, and magnets attached to the rotor that do. The coils are connected to some electronics that turn them on and off at the correct times. In the simplest configuration we could just turn the coils on and off in the right order (Figure 5-10) at a certain frequency and the rotor would be attracted to each in turn, running at a fixed speed.

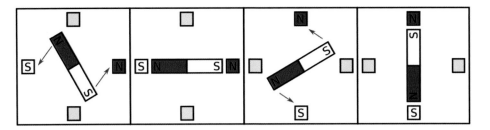

Figure 5-10 *How a brushless DC motor rotates by changing the magnetic field on its coils*

However, if the motor doesn't always run at a fixed speed the rotor may not be where the motor controller expects it to be. In this case the motor controller would turn coils on at the wrong time, and the motor would *stall*.

Imagine a modern cordless electric drill with a brushless motor. When running freely, the manufacturer knows the mass of every component in the drill and can calculate exactly how fast the motor will be able to start and stop. However, if you then start to drill a hole, the speed of the motor can vary wildly due to external factors. Without feedback the motor would stall.

To avoid this, most brushless motors incorporate some kind of rotation sensor. There is either an external sensor, or some circuitry to check the inductance of undriven coils (as this changes depending on the rotation of the rotor).

Sensors

The type of the sensor used often impacts how well the motor is able to restart. For instance, an electric car may have a very good sensor, allowing it to start moving smoothly regardless of the force applied to it. The kind of brushless motor that you'll find in a computer's cooling fan has only a sensor that detects when it has made a complete rotation (not its absolute position). If you purposefully slow it down and stop it, the motor will only be able to restart properly when there is no force applied to it.

Experiment 3: Stepper Motor

In some cases, there will be a known load attached to a brushless motor, or the motor can be made powerful enough to overcome the highest expected load. When this is the case, the motor controller can make a better guess about which coil it should turn on at what time, and you no longer need the rotation sensor.

We call this a stepper motor, because each time the motor controller changes the coil that is powered, the motor *steps* around.

A good example of this is the majority of consumer 3D printers. In 3D printers there is usually some kind of print head on a gantry that is moved around by motors. The mass of this gantry (and so how fast it can be accelerated and deaccelerated) is known by the manufacturer, and they can program this into the 3D printer.

In early (or home-built) 3D printers, this calibration was often left to the user, and it was easy to get it wrong as you were balancing making your printer fast with the accuracy of the motor. If the motor can't accelerate fast enough and loses just one step, your whole 3D print will have a *step* in it where the top half is offset from the bottom half!

You can make a stepper motor yourself with this hardware:

- 2x neodynium magnets
- 10cm cellophane tape or masking tape
- 2x paperclips
- 2x small nails
- A block of wood and two small screws
- A cork from a bottle of wine
- Two iron nails
- 2x 5m lengths of thin single-core insulated wire
- Four AA batteries in a battery holder

Let's create the motor:

1. Tape your two neodynium magnets to opposite sides of the cork. Make sure one has the north pole facing outward, and the other has the south pole facing out (if you hold a third magnet near the cork, one side of the cork should attract it, and the other should repel it).

2. Push one of the two small nails into each end of the cork, creating an axle for it to rotate on. You've now finished your rotor:

3. Next, make a stand for the rotor. Half-unfold each paperclip, and refold it into a Y shape, to give the nails of the rotor something to rest in.

4. Fold the other end of each paperclip tighter, and screw each paperclip into each end of the wooden block, so that you can rest the rotor into the Y of each paperclip.

5. Now we'll make the two electromagnets we need. Hammer the two iron nails partway into the wooden block, such that their tops are roughly at a 45-degree angle relative to the rotor:

6. Strip both ends of each piece of the single-core wire (strip slightly more off one end than the other). Leave 10cm of one end and then start winding it around one of the nails until you have 10cm left at the other end. Do the same for the other bit of wire on the other nail.

7. Place the rotor on top of the paperclips:

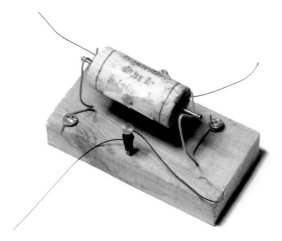

You may need to re-bend the clips a bit to get everything aligned nicely.

And we're done! Time to test the motor.

To test the motor:

1. Rotate the rotor so that one of the magnets on it is near one of the coils, but not completely facing it.

2. Take the two wires coming from that coil, and hold them across the battery pack for a second (put negative on the end that you stripped more off of). You should see the rotor move slightly. If the magnet moves away from the coil, rotate the rotor 180 degrees and try again!

3. You should now have one of the magnets facing the coil. Now connect the other coil across the battery for a second in the same way. The rotor should move 90 degrees so that one of the magnets now faces that coil.

4. Now connect the original coil back up, but this time with positive on the end that you stripped more off. The rotor should move again, but now the other magnet will be facing it, and you have managed to rotate the rotor a whole 180 degrees.

5. If you connect the other coil back up (and inverted too) the rotor will rotate to be a full 270 degrees from where you started, and connecting the original coil back up in the original pattern will have rotated the rotor a full 360 degrees!

If you could keep repeating this pattern (Coil 1, Coil 2, Coil 1 backwards, Coil 2 backwards, Coil 1, etc.) quickly enough, you could get the motor spinning.

Don't Want to Do This?

You can buy stepper motors with motor drivers very cheaply online, and many Arduino starter kits contain one.

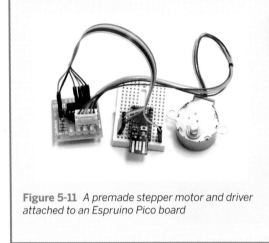

Figure 5-11 *A premade stepper motor and driver attached to an Espruino Pico board*

If you want to experiment with stepper motors but don't want to make your own, you could buy a stepper motor and wire it up as follows:

Motor driver	Espruino connection
-	GND
+	V_OUT
IN1	B1
IN2	A7
IN3	A6
IN4	A5

Experiment 4: Stepper Motor Control

Now that you've created the stepper motor from Experiment 3, you can use the microcontroller to automatically connect the coils up in the correct pattern quickly enough to spin the motor.

To make our motor move, we've had to put a lot of power (an amp or two) into our coils. The little microcontroller is only designed for turning small things on or off, and can supply only around 20mA (one-fiftieth of an amp!). To make it control our motor we'll need to use a motor driver IC to amplify the signals from it.

Don't Want to Do This?

In the following experiment, we're using a L293D motor driver IC. If you don't want to use one (or can't find one), you can get many premade motor-driver boards online. A board marked as a "Dual H-Bridge Motor Driver" should be exactly what you need. It should have:

- Input pins: ground, power, 4x inputs
- Output pins: 4x motor outputs

You'll need these electronics:

- A breadboard
- An Espruino board
- Patch wires for the breadboard
- An L293D motor driver IC

Follow these steps to control the motor:

1. Assemble the microcontroller board and motor driver as shown:

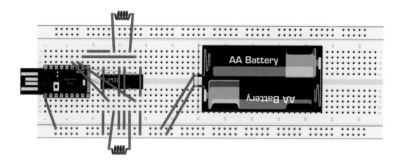

   This will have Espruino pins B1 and A7 connected to the first coil via the motor driver, and A6 and A5 connected to the second.

2. Connect the Espruino board to your computer as you did in Chapter 2, start up the Espruino Web IDE software, and click the *Connect* button in the top left.

3. You can now start to control your motor. As a first step, let's define the pins we're going to use. Type `var MTR = [B1,A7,A6,A5]` into the lefthand side of the Web IDE and press `Enter`. This will create a variable called `MTR` that is an array containing our four pins.

4. Now type `digitalWrite(MTR, 0b0000)`. This will make sure that all of the four wires are connected to GND, so all coils are turned off. We're giving `digitalWrite` a binary number as an argument, where each of the four digits after `0b` controls a pin. The first controls `B1`, the second `A7`, and so on.

5. When we wired up our circuit, we wired `B1` and `A7` to the first coil via the motor driver. This means we could turn our coil on in one direction by raising `B1` while keeping `A7` lowered. We can do this just by feeding the binary number `0b1000` into `digitalWrite`. Press `↑` to show the last command, then use `←` to step back-

wards to the first `0` character, and change it to a 1. Press `End` (or use the `→` to step to the end of the line) followed by `Enter`. The rotor should now move!

6. Next, it's a good idea to know how to turn the coil off to stop it from getting too warm! Press `↑` twice until the `digitalWrite(MTR, 0b0000)` command is shown again, and press `Enter` to execute it. This will connect all pins to ground again.

7. Next we can turn the other coil on. Do this using the same steps as before, but change the number to `0b0010` instead. This will set `A6` to 1, and `A5` to 0.

8. Then we need to turn the original coil on, but backwards. Just as we did manually, we'll just put positive voltage on the other pin. We used `0b1000` before, but this time we can use `0b0100`.

9. Now, turn the second coil on backwards with `0b0001`. The rotor should have rotated 270 degrees now.

10. To rotate it a full 360 degrees you can call up the previous command. Just press `↑` until you get to `digitalWrite(MTR, 0b1000);` and then press `Enter` again.

11. So altogether we've executed the following commands. You can keep pressing `↑` four times and `Enter`, executing each command in turn manually to turn your stepper motor!

```
digitalWrite(MTR, 0b1000);
digitalWrite(MTR, 0b0010);
digitalWrite(MTR, 0b0100);
digitalWrite(MTR, 0b0001);
```

The rotor will be attracted to each coil in turn:

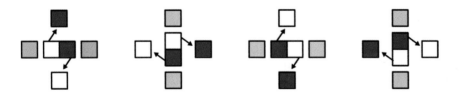

12. It would be great if we could get Espruino to do this automatically, but we can't just execute those commands in sequence because everything would happen too fast. We need to execute each one with a delay between. To start off, we'll store all our steps in an array and will make a function called `doStep` that steps through them. Copy and paste this into the lefthand side of the Web IDE:

```
// the 4 steps we're doing
var steps = [0b1000, 0b0010, 0b0100, 0b0001];
// the step we're going to output next
var step = 0;

function doStep() {
```

```
// output the step from the array
digitalWrite(MTR, steps[step]);
// move on to the next step
step++;
// but if there are no more steps, we must go to the beginning
if (step >= steps.length) step = 0;
}
```

13. By itself this won't do anything, but if you now type `doStep()` repeatedly the motor will turn 90 degrees each time.

14. To do this automatically we'll use Espruino's `setInterval` function to call `doStep` twice a second. It takes a number in milliseconds, so we need to use 500. Type `setInterval(doStep, 500)` and press `Enter`. Your motor should now start spinning!

15. Now that we've got Espruino calling our function, we can have the board call the function more often in order to rotate our motor faster. When we typed `setInterval` before, it returned the number `1`. This is the number of the interval that we just created, and we can use it to tell Espruino to change that interval. Type `changeInterval(1, 400)`. This will change the interval to be called every 400ms (2.5 times a second), and the motor should rotate a bit faster.

16. Try slowly decreasing the number you put into `changeInterval` and see how fast you can get the motor spinning! If you set it too high you might find that the motor will stop rotating properly and will just start shaking. You'll have to go back to a much lower speed to get it working again.

    You'll probably find that you can get the motor going fastest if you slowly increase the speed, rather than trying to start at a high speed. This is because the motor has inertia, so it can't suddenly change speed. Because the microcontroller doesn't know that the motor isn't at the correct speed, it'll start to put the wrong coil on at the wrong time, and will no longer be able to keep the motor moving.

17. At some point you'll probably want to stop your motor. To do this, you need to do two things: you want to stop changing which coils are on by stopping the interval, but you also want to make sure that you don't leave one of the coils on. You can do this by typing two commands back to back: `clearInterval();digitalWrite(MTR, 0b0000);`.

`clearInterval()` will clear all active intervals and timeouts. However, maybe you had multiple motors and only want to stop one! In this case you can pass in the interval's number just as you did for `changeInterval`. For instance, you can type `clearInterval(1)` instead.

---

18. If you want to wrap all of this into a handy bit of code that you can use to control your motor, you could do something like this:

```
// our pins
var MTR = [B1,A7,A6,A5];
// the 4 steps we're doing
var steps = [0b1000, 0b0010, 0b0100, 0b0001];
// the step we're going to output next
var step = 0;
// the interval we'll be using
var interval;

function start(rpm) {
  // just in case!
  stop();
  // start our interval
  interval = setInterval(function() {
    // output the step from the array
    digitalWrite(MTR, steps[step]);
    // move on to the next step
    step++;
    // but if there are no more steps, we must go to the beginning
    if (step >= steps.length) step = 0;
  }, 60000/rpm*(steps.length));
  /* revs per minute = 60*1000 seconds, but we have to call
  this function once for each step to make a complete revolution */
}

function stop() {
  // remove interval if there was one
  if (interval)
    clearInterval(interval);
  interval = undefined;
  // turn off coils
  digitalWrite(MTR, 0b0000);
}
```

19. You can enter this code on the lefthand side (or you can paste it into the right-hand side and click the *Upload* button), and can then type simple commands to start the motor. For instance, you can use `start(60)` to start the motor at 60rpm, then `start(90)` to speed it up, and finally `stop()` to stop it.

Experiment 5: More Stepper Motor Control

You might have wondered why we conveniently stored `steps` as an array. Sometimes there are better ways of turning the coils on and off to make the motor move:

1. Start the motor turning slowly with `start(30)`.

At the moment we've only ever got one coil on at a time. What would happen if we always had two coils on? The rotor would then be attracted by both coils and so would be able to produce more rotational force. However, you don't get this for free: two coils on means twice as much power is being used!

2. To make this happen, we can just replace the steps array by typing `steps = [0b1010, 0b0110, 0b0101, 0b1001];` on the lefthand side of the IDE.

3. But if instead we want our motor to move more smoothly (or maybe more accurately!) we can use a different step pattern that alternates between turning one coil or two coils on. In this case, when one coil is on, the magnet on the rotor will be attracted to that one coil, but if two coils are on then the magnet will be attracted to a position *between the two coils*:

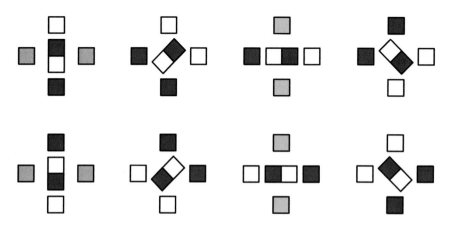

In terms of positioning, it's a bit like having twice as many coils! Try entering the following:

```
steps = [0b1000, 0b1010, 0b0010, 0b0110, 0b0100, 0b0101, 0b0001, 0b1001]
```

There are now twice as many steps, so the motor will only run at half the speed. However, it will be able to rotate much more smoothly at lower speeds.

Of course there's no reason why you can't do this kind of thing to an even greater extent. You could have one coil on at 100%, while the other was on at only 50%. That would move the rotor to roughly 1/3 of the way between the two coils. This is called *microstepping*, but we won't cover it here.

In this chapter we've controlled a stepper motor from first principles. In reality, it will probably be easier to use a library of code that somebody else has made.

Espruino contains a Stepper Motor driver library (http://www.espruino.com/StepperMotor) that can easily be used to perform normal stepper motor operations.

Stroboscope Tachometer

§

In this chapter we're going to explore what we can do with careful timing when we have some information about the rotation of an object.

To save us a bit of work, we'll use a premade motor and sensor combination. Probably the cheapest and most readily available source of these is the computer industry. Since computer fans need to be highly reliable (running 24 hours a day, 7 days a week, for years on end), brushed motors couldn't easily be used because the brushes would wear out. Instead, computer fans use brushless motors.

To help with efficiency and to lower noise, it's often useful to run these fans slowly when the computer doesn't need to dissipate as much heat. To do that reliably, the computer needs to know how quickly the fan is running. This means that many computer fans have an extra wire that pulses once per revolution. It allows the computer to detect the speed of the fan and to then adjust the voltage on the fan to keep it constant, or to produce an audible warning if the fan is stationary.

Finding a Fan

The easiest way to find a fan is just to buy one. If you search online for 80mm fan you'll find a huge array of fans to choose from. All you care about is that they're 12 volt, black, and have a wire with three or ideally four pins. You'll find that almost any computer parts store will stock them.

While you could buy a fan, you might have an old desktop computer that is due to be thrown out. In that case you will almost certainly find a fan inside if you open it. Again, you're looking for a fan that has three or four wires. Figure 6-1 is an example of a 4-wire fan from an Intel CPU. Figure 6-2 shows a 3-wire fan from a PC heat sink, and Figure 6-3 shows the 3- and 4-wire plugs side by side.

Figure 6-1 *A 4-wire fan from an Intel CPU*

Figure 6-2 *A 60mm 3-wire fan*

Figure 6-3 *3- and 4-wire plugs*

Experiment 6: Detecting Speed

You'll need the following:

- A breadboard
- An Espruino board
- 3 patch wires for the breadboard
- 3 or 4 wire fans (previously described)

Here are the steps to detecting speed:

1. Put the Espruino into the breadboard:

Connect the black wire of the fan to `GND`, the red wire to `5V`, and the yellow/white wire to pin `A8`. If you have a fourth blue wire, connect this to `B7`. Note that these fans are 12V and we're only running them off of 5V, so they will turn quite slowly, but that's good!

2. Now plug in the Espruino board, and connect with the Espruino Web IDE. If you have a 3-wire fan it should start spinning immediately, but a 4-wire fan might not.

3. If you have a 4-wire fan, the fourth (blue) wire is the PWM wire. This stands for Pulse Width Modulation, and it allows the computer to which it is connected to control the fan's speed based on how much of the time the signal is on versus how much it is off. For the moment we want our fan running as fast as possible, so we'll just turn that pin on by typing `digitalWrite(B7, 1)`.

4. Next, let's just store our sensor's pin in a variable with `var SENSE = A8;`.

5. The fan's sensor output is what's called Open Drain. When the fan blades are in a certain position, the output will be shorted to ground. When the blades are anywhere else in their spin, the output will just be left disconnected. This means that if we just measure the output, we won't see anything at all. We need to add a *pull-up resistor* to pull the voltage on the sense output up to a high level so that we can detect when the output isn't shorted to GND. We could do this with a resistor, but handily modern microcontroller chips have resistors built in that can easily be applied in software. Simply type `pinMode(SENSE, "input_pullup")` to turn on the internal pullup resistor.

Now, it would be helpful to see what's happening on the pin itself. We could just display the state on the Espruino board's LED. To do that, we'll want to change the state of an LED when the pin changes state. We'll do that using setWatch (which calls the supplied function whenever a pin changes state).

6. Just type the following in the lefthand side:

```
function onChanged(e) {
   digitalWrite(LED1, e.state);
}
setWatch(onChanged, SENSE, { edge:"both", repeat:true });
```

7. You should now find that the LED is lit. You might even see it blinking on and off. Try slowing down the fan with your finger, and you should see the blinking slow down as well. In fact, if you stop the fan and rotate it yourself, you should see the points of rotation at which the output is on, and the points at which it's off.

8. By itself that's not very useful, but we could change our onChanged function so that it was able to measure how many times the fan has rotated per second. Try the following:

```
var counter = 0;
function onChanged(e) {
   counter++;
   digitalWrite(LED1, e.state);
}
```

9. Entering the new code has caused the onChanged function to be overwritten. You don't have to call setWatch again; it's now executing the new code instead. Type counter and press Enter, and the value of the counter should be displayed.

10. Press ↑ and Enter again to print the value of counter again. It should be increasing each time the fan's sense output changes state.

11. We can now print the counter's value and set it to zero every second with the following code. That'll give us an idea what the speed of the fan is:

```
function onSecond(e) {
   console.log(counter);
   counter=0;
}
setInterval(onSecond, 1000);
```

12. Normally it should be printing a relatively constant value, but if you slow the fan down with your finger, you should see that the value reported drops off. We could even use this to output some kind of warning if the fan was running too slowly. Try the following:

```
function onSecond(e) {
   // light an LED if too slow
   digitalWrite(LED2, counter < 30);
   // output the revs per minute (60 seconds, but rising *and* falling)
   console.log(counter * 60 / 2);
```

```
    counter=0;
  }
```

13. Now, if you slow the fan down, the second (green) LED will light up. Once the fan is spinning again the LED will go out, but as we're only checking once a second it will always take a second to react to changes. It would be better if we could work out the speed using the time between pulses. We can do that by changing the onChanged function so that when it is called, it compares the current time with the last time it was called and uses that to work out the RPM:

```
// the last speed we calculated
var rpm;

function onChanged(e) {
  if (e.state) {
    // when the pin changes state to be high
    var timeDiff = e.time - lastPulseTime;
    lastPulseTime = e.time;
    rpm = 60 / timeDiff;
    digitalWrite(LED2, rpm < 900);
  }
  counter++;
  digitalWrite(LED1, e.state);
}

function onSecond(e) {
  // only light the LED if it's been a whole second without any move
ment   // (we must be stationary!)
  if (counter==0) digitalWrite(LED2, 1);
  counter=0;
  console.log(rpm);
}
```

 We've added an if (counter==0) ... check to the onSecond function. Without this, if you stop the fan from turning, there will be no changes in the signal from the fan and so our warning light wouldn't light up!

So now we've got our fan spinning, and we're able to see exactly how fast the signal is changing and work out the fan's RPM.

But is it correct? How can we actually tell?

One way is to use a stroboscope. If we create a short pulse of light when we think the fan has gone through one rotation, it should illuminate the fan at just that point.

Stroboscopes

A stroboscope is a device that makes an object that moves in a repetitive way appear to move slowly or even to be stationary.

It does this by hiding the object when it is everywhere other than where you want to see it. What would have been a blur now becomes a (darker) picture of the object in the correct position.

Joseph Plateau made the first stroboscope in 1832, using a disc with rotating slits. As each slit flew past a light source, it lit up a scene for a split second (much like a zoetrope). In 1917 Etienne Oehmichen patented the electric stroboscope, which used an electrically generated flash of light to illuminate the subject, rather than a slit.

Experiment 7: Stroboscope

You'll need the following:

- Everything as set up for "Experiment 6: Detecting Speed"
- A small white sticky label

Follow these steps to use the stroboscope:

1. To make sure you've reset your Espruino board (if you were using it for the last experiment), type `reset()` on the lefthand side and press `Enter`. The board will reset, and the text that was printing once a second will stop.

2. Write your initial on the sticky white label, and put the label on the outer edge of one of the fan blades, so you can easily see it.

3. Now we want to pulse an LED whenever the signal from the fan changes from a 0 to 1. For this we'll use `setWatch` as we did before, but will set the edge to `rising` rather than `both`. We can also use `digitalPulse` to pulse the onboard LED for a short period of time. We're using LED2 because the green LED is slightly brighter than the red one!

```
var SENSE = A8;
pinMode(SENSE, "input_pullup");
function onChanged(e) {
  digitalPulse(LED2, 1 , 2/*ms*/);
}
setWatch(onChanged, SENSE, { edge:"rising", repeat:true });
```

The green LED should start flickering on and off.

You'll have to turn off your lights and put your fan very close to the Espruino board, but you should see something pretty amazing: your moving fan will appear to have stopped moving! You'll still hear the motor whir, and you'll still feel the breeze from the fan blades, but the image of the fan blades will appear to be stationary, as in Figure 6-4.

Figure 6-4 *The label highlighted by the strobe of the green LED*

What is happening? The green LED is flashing at exactly the same time that the fan motor is completing one revolution. When the scene lights up, you see the blades of the fan in the exact same spot they were a fraction of a second before. By keeping the flash of light short, and the rest of the room dark, you'll appear to see the blades in the same spot all the time, even though they're still spinning.

You may find that you can see two distinct copies of your sticker, 180 degrees out from each other. This is because the fan's sensor actually changes state four times per revolution (high - low - high - low), and it means that our previous RPM measurements were twice as fast as they should have been!

So what can we do if we just want the sticker to be highlighted once per revolution? We can count each time the signal changes, and can only pulse the LED when our counter is odd rather than even.

A fast and simple way to do this on a computer is to take advantage of the fact that it uses binary arithmetic. As everything works in base 2, the bottom digit (or bit) will be 1 if the value is odd, and 0 if it is even. All we have to do to get it is to use the binary & operator:

```
var counter = 0;
function onChanged(e) {
  counter++;
  if (counter&1)
    digitalPulse(LED2, 1 , 2/*ms*/);
}
```

Now, the sticker should only be highlighted once. You might even be able to read your initial on the fan's spinning label.

Now that we're only flashing the LED once per revolution, it'll be getting really hard to see. Ideally we'd have something that we could see more easily for our next experiments. Let's try and use a brighter light.

Experiment 8: Brighter Stroboscope

You'll need the following:

- Everything as set up for Experiment 7
- A breadboard
- An Espruino board
- A P36NF06L FET (alternatives in Appendix A)
- An ultra-bright LED and a 100 Ohm resistor

Complete the following steps:

1. Connect everything as shown in the diagram, leaving the fan connected:

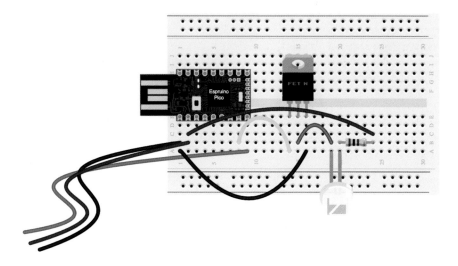

2. Arrange the LED such that it is pointing toward the fan blades.

3. Now, all you need to do is change the pin in your code from LED1 to the pin that the light is on (B6) and re-upload it. Reset the Espruino with reset() and upload the following:

```
var SENSE = A8;
var LIGHT = B6;
var counter = 0;

function onChanged(e) {
  counter++;
  if (counter&1)
    digitalPulse(LIGHT, 1 , 2/*ms*/);
}

digitalWrite(LIGHT, 0);
pinMode(SENSE, "input_pullup");
setWatch(onChanged, SENSE, { edge:"rising", repeat:true });
```

If you point your LED at the fan, you should now be able to see your sticker high-lighted much more clearly.

John Logie Baird's TV   7

You might have noticed when you did "Experiment 8: Brighter Stroboscope" that the label didn't appear entirely solid. It had some motion blur. This happens because the light didn't just flash on and off in an instant. It turned on, and turned off two thousandths of a second later. However, in those thousandths of a second, the fan blade had actually moved a noticable amount.

This is exactly the same reason you get motion blur in a still picture, and we can play with the phenomenon to produce some really interesting effects!

Experiment 9: Persistence of Vision

You'll need:

- Everything as it was set up for "Experiment 8: Brighter Stroboscope"

Now follow these steps:

1. Upload the following code (this is from "Experiment 8: Brighter Stroboscope", but we've just moved the pulse length argument from `digitalPulse` into a variable called `pulses`):

```
var SENSE = A8;
var LIGHT = B6;
var counter = 0;
var pulses = 2;/*ms*/

function onChanged(e) {
  counter++;
  if (counter&1)
    digitalPulse(LIGHT, 1 , pulses);
}
```

```
digitalWrite(LIGHT, 0);
pinMode(SENSE, "input_pullup");
setWatch(onChanged, SENSE, { edge:"rising", repeat:true });
```

If you only saw a single sticker in Experiment 7 then you should leave the `if (counter&1)` line out for this, and for subsequent experiments.

You should now be able to see the dot highlighted by the strobe, exactly as before.

2. Type `pulses = 0.5`. This will reduce the pulse length to 0.5ms from 2ms, and the dot should appear a lot sharper, with less blur (but a little more dim). It's like making the shutter speed faster in your camera.

   You can go the other way, too. `pulses = 5` will make your dot into a big long smear.

You can experiment with different values, but be careful! If you make the pulse take longer than the fan takes to rotate, more pulses will keep getting queued up until Espruino becomes unresponsive.

`digitalPulse` can do more than just pulse the LED once. If you give it an array as an argument then you can specify not just the time the output will stay high for, but how long it will be low for after that, how long it'll be high after that, and so on.

3. Try `pulses = [0.5, 5, 0.5]`. You should now see two copies of your label, because there are two pulses separated by 2ms.

4. Now try `pulses = [0.5, 2, 0.5, 2, 0.5]`. You should see three distinct copies of your label. But now, what happens if you slow the fan down with your finger? The pulses will get closer together.

   If we want to make sure the pulses appear at certain positions we'll want to make their spacing dependent on the time it took to do a revolution. In Experiment 6 we worked out how long a pulse from the sensor took by looking at `e.time` in our watch function, so let's add that code here and use it to multiply every value in the `pulses` array.

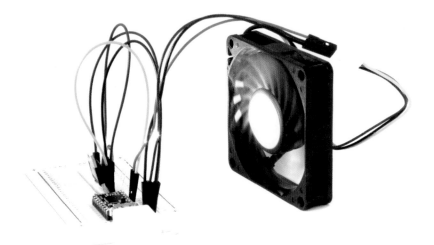

5. Type the following code in the Web IDE editor:

```
var lastPulseTime;

function onChanged(e) {
  counter++;
  if (counter&1) {
    var d = e.time - lastPulseTime;
    lastPulseTime = e.time;
    var p = pulses.map(function(t) { return t*d; });
    digitalPulse(LIGHT, 1 , p);
  }
}
```

The preceding code is using the JavaScript array's `map` *function to multiply every element of the array by the variable* `d`. *The* `map` *function creates a new array by calling the supplied argument supplied on each element.*

For example, `arr.map(fn)` *is the same as* `[fn(arr[0]), fn(arr[1]), fn(arr[2]), .... fn(arr[arr.length-1])]`.

Chances are you won't be able to see very much, because we're multiplying the contents of the `pulses` array (which should be in milliseconds) by the time between sensor pulses (which was in seconds). This means that to light up for a full revolution we'd need a value of 1000 in our `pulses` array, but we've got `0.5` and `2`.

6. Try setting the `pulses` variable to something a bit more simple with **pulses = [5, 50, 5, 50, 5]**.

   You should now see the three disinct copies of the label that you had before. However, if you slow down the fan with your finger the pulses will now stay roughly the same distance apart.

   It's even possible to dynamically change the timings to make a simple animation. We can use a sine wave to move the middle point slowly backward and forward:

   ```
   function animate() {
     var l = 50 + 40*Math.sin(getTime());
     pulses = [5, l, 5, 100-l, 5];
   }
   setInterval(animate,100)
   ```

   While having a single pulse will make the fan look like it's stopped, we can actually do better. By pulsing once for each fan blade we can show all the fan blades overlapped.

7. First, we'll want to remove our animation. Type `clearInterval()` to stop it, and then type **pulses = [5]** to go back to a single pulse.

8. Next, stop the fan and count how many fan blades it has. The one I'm using here has nine, but most fans have seven so that's what I'll use for the examples.

   Now, we want to work out what the spacing between pulses should be. We know we need 1000 for a full revolution, so we just subtract the amount of time we have to spend pulsing the light once for each fan blade, and then we divide by the number of fan blades.

9. We might as well save this information into a variable by typing `var blades = 7;` and `var t = (1000 - blades*5) / blades;`.

10. Now we just want to set up the `pulses` array:

    ```
    pulses = [5];
    for (var i=0;i<blades;i++) pulses.push(t, 5);
    ```

Now you should be able to see one fan blade that's a composite of the other blades.

Experiment 10: John Logie Baird's TV

John Logie Baird's TV

In 1884 Paul Julius Gottlieb Nipkow patented the Nipkow disc. This was a disc with a series of holes in it, placed in a spiral. As the disc rotates, each hole scans out one line at a time. If you put a light behind the disc and turn it on and off at the right time, you can use the persistence of vision of your eye to scan out a complete picture.

While Nipkow never actually made anything with his patent, in the 1920s technology had advanced enough that John Logie Baird was able to use a Nipkow disc to make both a camera and television, which was used for a short time by the British Broadcast Corporation, among others.

In the end, fully electric televisions based on cathode ray tubes took over. They were quieter, higher resolution, and more reliable. There's still something very intriguing about an entirely mechanical television system!

In this experiment we'll make our own mechanical television in the style of John Logie Baird's. While we could get a better picture using a proper Nipkow disc (with holes in it), that would take quite a bit of time and effort. Instead of shining light through holes, we can use our fan setup from Experiments 8 and 9 and can reflect the light off white labels instead!

You'll need:

- Everything as it was set up for "Experiment 8: Brighter Stroboscope"

- Small white stickers (or white paint/Wite-Out)

  - Peel off the single sticker you'd put on the fan previously.

  - Add new stickers near the leading edge of each fan blade, each one a little bit nearer the middle than the last. You can use white paint or Wite-Out for this, but it's harder to reposition if you're not happy with it!

- Add the code we ended up with after "Experiment 9: Persistence of Vision":

```
var SENSE = A8;
var LIGHT = B6;
var counter = 0;

var blades = 7;
var t = 1000 / blades;
var pulses = [5];
for (var i=0;i<blades;i++) pulses.push(t-5, 5);

var lastPulseTime;

function onChanged(e) {
  counter++;
  if (counter&1) {
    var d = e.time - lastPulseTime;
    lastPulseTime = e.time;
    var p = pulses.map(function(t) { return t*d; });
    digitalPulse(LIGHT, 1, p);
  }
}

pinMode(SENSE, "input_pullup");
setWatch(onChanged, SENSE, { edge:"rising", repeat:true });
```

You should now see a straight line highlighted on the fan (made out of each of your stickers). This is because, as with "Experiment 9: Persistence of Vision", you're seeing the composite of all of the fan blades:

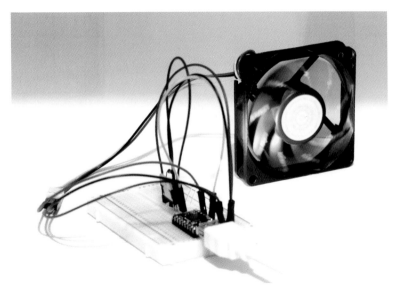

However, by turning the light on and off for different amounts of time, we can make different patterns.

- Try the following code, but add or remove 5, t, lines such that you have one line per fan blade:

```
pulses = [
50, t-50,
5, t-5,
5, t-5,
5, t-5,
5, t-5,
5, t-5,
50];
```

On most fan blades you'll see an assortment of lines, but on one blade you should see a clear C shape:

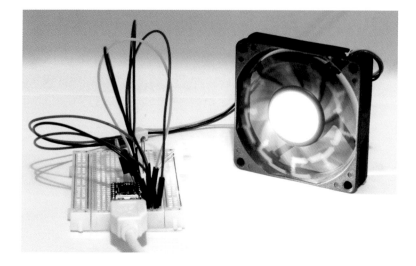

This works because each array element in `pulses` contains a list of times that the LED light should be on and off for. Even elements (array indices start from 0, so this starts from the first element in the array) represent the amount of time that the light should be on, and odd elements are times that the light should be off.

These times are relative, so we multiply them by the time between rotations of the fan. The time between rotations is in seconds and `digital Pulse` accepts times in milliseconds, so all of the elements in the `pulses` array are in thousandths of a rotation of the fan blade.

We calculated `t` as the relative time taken for one fan blade to rotate, so each line of the image we want to write should add up to `t`. Hence for the very top of the `C` where we want the light to be on a long time we use `50`, and then we need to turn the light off for `t-50` to ensure that when we next turn it on, it's right at the start of the next line.

In the middle we use `5` and `t-5` for a short pulse of light, and finally at the end we don't need to make up a complete line so we only supply one value (`50`), which is the time we want to turn the light on in order to make the bottom edge of the `C`.

- In fact, by timing the pulses correctly you can output whatever image you want. Try the following, which will output a square:

```
var pulses = [
50, t-50,
5, 40, 5, t-50,
5, 40, 5, t-50,
5, 40, 5, t-50,
5, 40, 5, t-50,
```

```
5, 40, 5, t-50,
50];
```

In fact you can do whatever you want here, as long as each line has an even number of elements (so the light turns off at the end of each line) and adds up to t.

Manually setting the pulse widths this way is pretty frustrating, so you can create a function to automate this. For example the following will let you draw your own images just by passing in an array of strings with X or a space to signify the light being on or off:

```
function toPulses(img) {
  pulses = [];
  // We're pulsing high at the start of the image
  var lastPixel = true;
  // the time the light will be on or off for
  var time = 0;
  // iterate over the image
  for (var y in img) {
    var line = img[y];
    for (var x in line) {
      var pixel = line[x]!=" ";
      if (pixel!=lastPixel) {
        // if this pixel is different, output the time
        pulses.push(time);
        time = 0;
        lastPixel = pixel;
      }
      time += 5;
    }
```

```
    // end of line, turn off
    if (lastPixel) {
      pulses.push(time);
      time = 0;
      lastPixel = false;
    }
    time += t - line.length*5;
  }
}

toPulses([
  "XXXXX",
  "X   X",
  "X   X",
  "XXXXX",
  "X   X",
  "X   X",
  "XXXXX"]);

// or...

toPulses([
  "XX    XX",
  "XX    XX",
  "        ",
  "X      X",
  " X    X ",
  "  XXXX  "]);
```

And that's it! You've made your own version of one of the earliest televisions using just an old fan, a light, and a microcontroller!

Want to experiment more? You could use `setInterval` to call `toPulses` with a series of different pictures in order to make an animation!

PART III

Electromechanics

Making things move!

Now that we've learned a bit about motors, let's try to use them to do something useful in the real world.

Make a Simple Robot

One of the most rewarding ways to start using control systems is to make an autonomous robot. Even very simple rules can create seemingly intelligent behavior.

While we made our own motor in the last chapter, to get started quickly here we're going to use a premade servo motor that contains a motor, gearbox, and the drive electronics. Probably the best source of these is remote-control cars, planes, and helicopters. All of these models need small, lightweight actuators that can move to a specific location on demand.

Servo Motors

In Chapter 5, we looked at stepper motors (which we control by turning electromagnets on and off in sequence), as well as the PC fan, which had a sensor that told us when the fan blades had rotated one revolution. With both of these motors we had some idea of where they were and how fast they were moving.

In many cases, however, you need an actuator that can be rotated to an absolute position.

Servo motors contain a motor connected to a position sensor and some control electronics. The control electronics can then turn the motor clockwise or anticlockwise until it reaches the desired position. Unlike a stepper motor, if some external force moves the motor away from the desired position, the control electronics can detect this and move the motor back.

RC servo motors generally use a standard brushed motor connected to a gearbox, which is in turn connected to a potentiometer (Figure 8-1). The potentiometer (or variable resistor) changes its resistance based on its rotation, and some simple control circuitry compares the new resistance with the expected resistance and moves the motor accordingly.

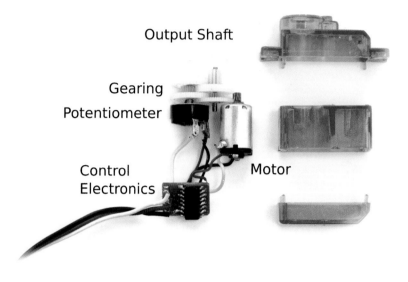

Figure 8-1 *The component parts of a servo motor*

However, because of the way these potentiometers are built, they can only be turned around a certain amount of times before the internal wiper hits an end stop. The cheap potentiometers used in most model servo motors can only be rotated around 270 degrees, so most servo motors are restricted to this or less.

These motors take a digital input in the form of a square wave. Every time the motor's control electronics get a square wave, they compare the length of that wave (the amount of time the signal is high) with the potentiometer's resistance, and move the motor slightly clockwise or anticlockwise. A signal length of 1.5 milliseconds (ms) moves the motor so that the potentiometer is in the middle. A signal length of 1.0ms moves the motor to one end of its range, and 2.0ms moves it to the other end. RC servo motor controllers are usually stupid, and it is possible to give the motors pulse lengths outside of the 1–2ms range, which will cause them to move a bit further, but that could also cause damage to the motor.

Most servo motors have an output shaft that can only rotate from 0–180 or 0–270 degrees because of their potentiometer. There is, however, a special kind of motor called a *continuous rotation* servo. These motors have the potentiometer disconnected, so they can rotate as many times as is needed (although not to any absolute position). While at first glance this might seem pointless, it does give us a great source of easily usable motors for robots!

Making Do

Don't have or can't find a servo motor? No problem! Just try to find some motors and gearboxes (perhaps from a premade three-wheeled robot). You can then use the L293D motor driver or motor driver boards mentioned in Chapter 5.

Just wire the motors up, and instead of using `analogWrite` in the following examples, set one output high and the other low to make the motor move one way, and the opposite to make it reverse.

Experiment 11: Try Out a Servo Motor

You'll need:

- An Espruino board
- Breadboard
- 3 patch wires
- A servo motor (see Appendix A for more information)

Follow these steps to test your servo motor out:

1. First take a look at your servo motor. It should have three wires coming out of it. They should be either black, red, and white, or in some cases brown, red, and yellow. They are connected as follows:

Color 1	Color 2	Connection
Black	Brown	GND
Red	Red	Power
White	Yellow	Signal

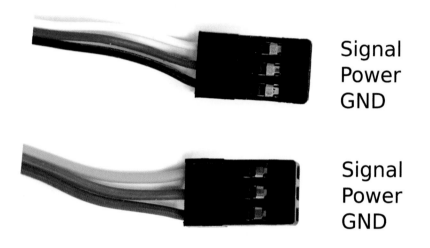

Signal
Power
GND

Signal
Power
GND

The remote-control model servo motors we are using here were originally designed to run off 4x NiCd AA batteries, so they expect a voltage around 4.8v. Luckily the Espruino boards contain a protection diode which drops the voltage available to a nearly perfect 4.7v.

2. Connect the wires up: GND to GND, Power to Espruino's 5V/VBat pin, and Signal to any data pin on the Espruino. Let's use pin `B3`.

3. Now plug the Espruino board into your computer, and connect with the Web IDE.

4. In the lefthand pane, type the following: `digitalPulse(B3, 1, 1.5)`.

   This will provide a single 1.5ms pulse to the motor, and for a fraction of a second the motor will move toward its middle point. If you're doing this on a continuous rotation servo, it shouldn't move at all because 1.5ms is the midpoint between forward and reverse.

5. Press ⬆ to select the previous command (`digitalPulse`) and press `Enter`. Do this repeatedly and you should find that the motor jerkily moves to the midpoint.

6. Now try moving the motor to one side, by sending a 1.0ms pulse: `digital Pulse(B3, 1, 1.0)`.

   You'll need to use ⬆ and `Enter` again a few times, and you should see the motor move, and then finally stop at its final position. If you're using a continuous rotation servo, it will keep moving forward all the time.

7. Obviously it would be better if we could do this automatically. In Chapter 3 we saw how we could use `setInterval` to execute code every few milliseconds, and we can do this here:

```
// the position of the motor
var pos = 0;

// To be called every so often to tell the servo where to go
function updateServo() {
  digitalPulse(B3, 1, E.clip(1.5+pos, 1, 2));
}

// Now make sure we call the function every 50ms
setInterval(updateServo, 50);
```

The servo should now jump back to its home position (if it's not continuous rotation), but you can now update the value of the `pos` variable to move the servo around.

8. Try `pos = -0.5`. This will move the servo to one side of its travel, and use `pos = 0.5` to move the servo to the other side.

Note that in the preceding code we've got `E.clip(1.5+pos, 1, 2)`. This will make sure that the pulse width is never more than 2ms or less than 1ms, which will help to protect your motor from damage.

You can make the motor move slowly between two points by varying `pos` over time. For example, we'll use a sine wave to slowly make the servo motor oscillate.

9. Modify the `updateServo` function by re-entering:

```
function updateServo() {
  pos = Math.sin(getTime()) * 0.5;
  digitalPulse(B3, 1, E.clip(1.5+pos, 1, 2));
}
```

You should now see the servo moving smoothly from one end of travel to the other. If you've used a continuous rotation servo it will start moving one way and will then slow down, stop, and start moving in the other direction.

Software Versus Hardware

In the preceding code, we're using `digital Pulse`, which outputs a single pulse using Espruino's software-based timer. Espruino also contains hardware timers, which can generate a continuous square wave independently of the software.

To do this, you can use `analogWrite`. You need to specify a frequency and a duty cycle (the percentage of time that the square wave generated should be high versus low).

To generate the same pulse we're doing with `digitalPulse` (every 50ms), we need a

`1000ms / 50ms = 20 Hz` square wave, set with the correct duty cycle:

```
function updateServo() {
  pos = Math.sin(getTime()) * 0.5;
  var len = E.clip(1.5+pos, 1, 2);
  analogWrite(B3, len/50, {freq:20})
}
```

However, now you need only call the function when you want the servo to move. The pulses will be sent in the background even without your function being called.

Experiment 12: Make a Simple Robot

You'll need the following:

- An Espruino Pico board
- A breadboard (with double-sided tape on the rear)
- 6 patch wires
- 2x *9g* size, continuous rotation servo motors (see Appendix A for more information)
- One fizzy wine cork (or two short elastic bands if your servos have a circular head; *lobster bands* are perfect—see Appendix A)
- A USB power pack and Type A extension lead
- A paperclip
- An elastic band

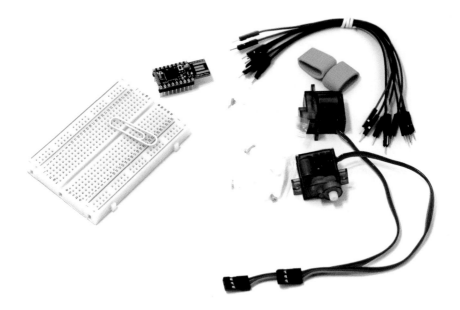

Follow these steps to make a robot:

1. Firstly, get your two continuous rotation servo motors, and peel the sticky labels off the side of them (they'll just get in the way).

2. If your servo motors came with a circular *plate*, push it on and screw it in. Otherwise use the cross-pattern plate.

3. If you have *lobster bands*, stretch them over the circular servo head:

Later on, these will be used as wheels for our robot:

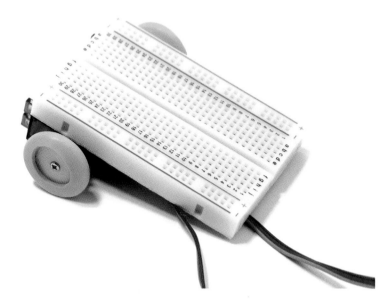

If you decided to use the champagne cork (or used the cross-pattern servo plate), carefully cut two thin (4mm) slices of cork off of the expanded end of the champagne cork. Use hot glue to glue them onto the servo plate.

4. Now, push the Pico into the breadboard with the connector sticking out of one side as far as possible.

5. Use patch wires to connect power and ground to both servo motors, and connect the lefthand servo (looking at the breadboard with the Pico at the rear) to pin B3 and the righthand one to B4:

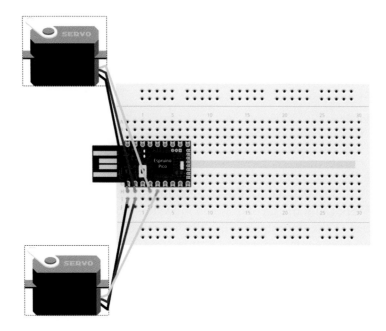

6. Now that the servo motors are connected, let's test them out! Use the Type-A USB extension cable to plug the Espruino Pico into your computer, open the Web IDE, and connect.

7. Copy and paste the following in the lefthand side:

```
var motors = [0,0];
function updateServos() {
  // Left
  digitalPulse(B3, 1, E.clip(1.5+motors[0], 1, 2));
  // Right
  digitalPulse(B4, 1, E.clip(1.5-motors[1], 1, 2));
```

```
}
setInterval(updateServos, 20);
```

At the start, nothing will happen because we're sending a signal to the servos that is telling them to stay in their middle state. You may still hear a slight ticking noise though!

Are Your Motors Moving?

Sometimes continuous rotation servo motors don't come very well calibrated in the factory. They often have a small adjustment in the bottom of the motor that can be moved until the motor no longer turns. If you have this, put a small screwdriver in the bottom of the servo and rotate the adjustment until the servo is no longer moving. Often you will only have to move the adjustment by a few degrees.

Don't worry if you don't have an adjustment, you can do it in software! Simply press ↑ until the declaration of the updateServos function is shown, and use the arrow keys to move the cursor to the text 1.5 for the servo that is moving. Try changing the value to 1.45, press ↓ to move to the end of the function, and press Enter to execute.

If this makes things worse, try the opposite: repeat the same steps and change the value to 1.55. Try different values until you find something that works.

You can now make the motors move by entering commands like motors = [1,0] and motors = [-1,-1]. The first element of the motors array should move the motor on the lefthand side of the robot, and the second should move the motor on the right. [1,0] will make the robot move to the right (by moving the left motor forward), and [-1,-1] will make the robot move backwards by reversing both motors.

8. Enter motors = [0,0] to stop them.

9. Now that the servos are working, it's time to assemble the rest of the robot. Unplug the robot from your computer.

10. Place the two servo motors in the rear corners of the breadboard with the wheels facing out to either side, and mark off the edge with a pen.

11. Get a sharp knife, cut along the sticky-back plastic, and then peel off the smaller half of the plastic:

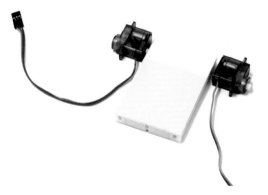

12. You can now stick both servo motors in place on the breadboard:

Now all that's needed is something for the third wheel of your robot. Unfortunately, tiny castor wheels are hard to get hold of, so we'll have to make do.

13. Take a paperclip, fold the middle out at 30 degrees, and then bend it into a curve:

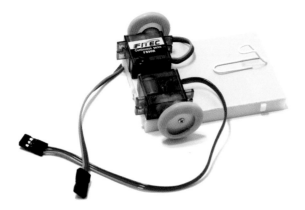

14. Peel off a small bit of sticky tape and attach the paperclip to the bottom of the breadboard:

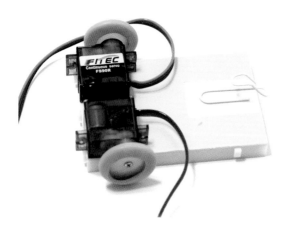

Place a bit of sticky tape over the top to hold it in place:

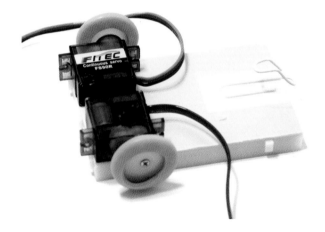

15. Now, carefully fold up your wiring and tidy it with an elastic band to keep it from touching the floor. It makes a big difference, as you can see:

Now that you've made your robot, it's time to make it do something!

Follow these steps to make your robot do something:

1. Connect the Espruino Pico to your computer and the Web IDE again.

2. You can now control the robot in the same way we did for "Experiment 11: Try Out a Servo Motor". Type `digitalPulse(B3, 1, 1);`. The left motor should now move a small amount in one direction.

3. Try `digitalPulse(B3, 1, 2);`. The left motor should now move slightly in the other direction, and you can type `digitalPulse(B4, 1, 2);` to move the right motor.

4. Entering `digitalPulse(B3, 1, 2);digitalPulse(B4, 1, 1);` should move the robot forward (or backward) a little. Note that because the motors are opposite each other we actually have to turn one clockwise and one anticlockwise in order to move forward.

5. Now that we've done that, we can work on making the robot move of its own accord. Enter the whole `updateServos` block of code in the Web IDE that we used earlier.

6. Now try typing `motors = [1,1];`, and then type `motors = [0,0]` to stop. The robot should have moved forward. If it moves backwards, no problem! Just re-enter the `updateServos` function, but swap around the `+` and `-` in the `E.clip` function calls.

   It would be nice if we could automatically move forward for a while, and could then do something else. We can do this using `setTimeout`.

---

7. Enter `motors = [1,1]; setTimeout(function() { motors = [0,0]; }, 500);`.

   This will move the robot forward for half a second (500ms) and will then stop it, but it's a bit long winded especially if we want to do multiple things.

   Instead, we could create a function that would make the robot do what we want. We will input a list of moves, and the robot will execute them one after the other:

   ```
   // the names of our movements
   var FWD = [1,1];
   var BACK = [0,0];
   var LEFT = [-1,1];
   var RIGHT = [1,-1];
   var STOP = [0,0];

   function go(moves) {
     // take the first command off our array
     var move = moves.shift();

     if (move) {
       // Move the motors
       motors = move;
       // Call ourselves again in half a second,
       // with the remaining list of moves
       setTimeout(go, 500, moves);
     } else
       motors = STOP;
   }
   ```

8. You can now try it with `go([FWD,LEFT,FWD,BACK])` and the little robot will move forward, turn left, move forward, then back again.

9. You can experiment with a list of commands that's as long as you want. If you want the robot not to move for a while, just add the command `STOP`.

   Of course during this time you've got the wire connected to your robot all the time. It would be nice if you could make the robot do things all by itself.

   The only input we have available to us at the moment is the button on the Pico board, so let's use that.

10. First, we want to create a *watch* (enter the following code). This is a function that will execute whenever an external input changes state. You can change the `go` command in the middle to whatever you want; the extra options on the end tell Espruino to call the function each time the button is pressed, but not when it is released. Debouncing stops the code from being executed twice in quick succession if the button physically *bounces* when it is pressed (in this case we only call our function if the button has stayed pressed for at least 50ms).

```
setWatch(function() {
  go([FWD,LEFT,FWD,BACK]);
}, BTN, {repeat:true, edge:"rising", debounce:50});
```

11. Now you can press the button, and the pattern of moves will be executed.

12. Wait until the robot has stopped moving, and then type `save()` into the lefthand side of the IDE.

    Your code is now written onto the Espruino Pico, and can run without a computer.

13. Unplug the Espruino from your computer, attach the USB power pack, and hold it on top of the robot with an elastic band:

14. Now press the button. The robot will move all by itself!

While this is quite fun, with no real inputs (apart from the button) it's difficult to do much with the robot. What we need to do is give it some senses!

Our final code is:

```
var motors = [0,0];
function updateServos() {
  // Left
  digitalPulse(B3, 1, E.clip(1.5+motors[0], 1, 2));
  // Right
  digitalPulse(B4, 1, E.clip(1.5-motors[1], 1, 2));
}
setInterval(updateServos, 20);

// the names of our movements
var FWD = [1,1];
var BACK = [0,0];
var LEFT = [-1,1];
```

```
var RIGHT = [1,-1];
var STOP = [0,0];

function go(moves) {
  // take the first command off our array
  var move = moves.shift();

  if (move) {
    // Move the motors
    motors = move;
    // Call ourselves again in half a second, with the remaining list of moves
    setTimeout(go, 500, moves);
  } else
    motors = STOP;
}

setWatch(function() {
  go([FWD,LEFT,FWD,BACK]);
}, BTN, {repeat:true, edge:"rising", debounce:50});
```

You can enter this on the righthand side of the IDE and click the *Upload* button if you want to.

Experiment 13: Following Light

You'll need:

- The robot from *"Experiment 12: Make a Simple Robot"*
- 2x light-dependent resistors (LDRs)
- 2x 10k resistors
- 5 patch wires
- A flashlight

Follow these steps:

1. Connect the components to the breadboard as shown, keeping the servo wiring as it was in *"Experiment 12: Make a Simple Robot"*:

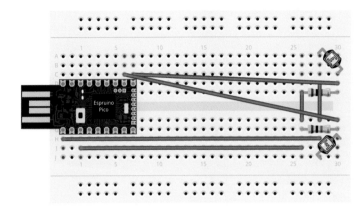

Once connected, your robot's LDRs should be pointing forward and to the sides:

This will connect your lefthand LDR to pin `A5`, and the righthand one to pin `A6`. The two resistors act as pullups, trying to move the voltage nearer the positive 3.3v they are connected to, while the LDRs pull the voltage down toward 0v.

2. Now connect the Espruino board to your computer again, and run the following command on the lefthand side: `analogRead(A5)`.

   This will give you a number between 0 and 1, which represents the voltage on pin `A5`.

3. Take a flashlight and shine it at the lefthand sensor, then run the command again.

   You should now have a second number, lower than the first. This is because LDRs lower their resistance when exposed to light. The lower resistance compared to the 10k pullup resistor causes the voltage to drop.

4. Without the flashlight, check the value from the righthand LDR with `analogRead(A6)`.

   If everything is connected well, and one side of the robot isn't facing anything bright, the value should be about the same.

   In order to make the robot move toward the light, we need to take a rough measurement of the values from the sensor when there is no light shining directly on the sensors.

5. Run the following code:

```
var darkValue = Math.min(analogRead(A5), analogRead(A6));`
```

This will take the lowest of the two light-sensor readings (corresponding to the one with the most light), and will save it to the `darkValue` variable.

6. Next, we'll use this to move the robot toward light in a really simple way. If the left sensor has more light (a lower value) than `darkValue`, we'll move the right motor forward, and vice versa.

When the light is to the left, the robot will move in that direction, but if it is straight ahead, both motors will turn, moving the robot forward!

7. Now enter the following code on the lefthand side of the IDE:

```
var motors = [0,0];
function updateServos() {
  var left = darkValue - analogRead(A5);
  if (left < 0) left = 0;
  var right = darkValue - analogRead(A6);
  if (right < 0) right = 0;
  motors[0] = right;
  motors[1] = left;

  digitalPulse(B3, 1, E.clip(1.5+motors[0], 1, 2));
  digitalPulse(B4, 1, E.clip(1.5-motors[1], 1, 2));
}
setInterval(updateServos, 20);
```

The robot may start moving a little of its own accord!

8. You can now type **save()**, plug the robot into the USB power pack, and start experimenting with it:

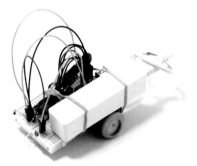

9. If you shine a flashlight at the robot now, it should start to move toward the light. If you turn the light off it'll stop moving, unless it is now facing something brighter than it was when it was calibrated!

You can easily extend this code to try different ways of controlling the robot:

- Use simple `if` statements to control in the robot in a more *digital* way.

- Make the robot avoid light instead of chasing it.

- When the robot first powers on, automatically set `darkValue` so that the robot can work in all kinds of different locations.

- By pointing the light sensors down, you can make your robot move around the edge of a dark shape. It can even find its was through a maze just by following one edge of it!

- By carefully positioning the sensors pointing downward you can make the robot follow a line—but that can be difficult to get working reliably unless you have a really thick line!

- You can also add other sensors like the HC-SR04 Ultrasonic distance sensor. This will let your robot move around of its own accord without hitting things!

Pen Plotter

9

In Chapter 8, we made our robot using continuous rotation servo motors, but the most common type of servo motor can only rotate by around 270 degrees. While these motors can't be used for robot wheels, their ability to repeatedly move to an absolute position makes them useful for all kinds of other things. Here we're going to use them to control the position of a pencil so that we can draw shapes.

Experiment 14: Pen Plotter

You'll need:

- A small corkboard (roughly 30cm x 40cm)
- 3x *9g* RC servo motors (*not* continuous rotation) with their servo plates
- One servo motor extension wire
- 9x patch leads
- Breadboard
- Espruino Pico
- A 47uF, 6v (or higher) capacitor (optional)
- An old pencil with soft lead (2B or softer is great—normal HB will work but is very faint)
- Two wooden chopsticks
- A hot glue gun
- Sticky tape
- A 2-inch square of thick noncorrugated cardboard

- 1 meter of thin string (kite string is perfect)

Here's how to assemble the pen plotter:

1. First, remove the stickers from the sides of your servos, and pick out the longest servo plates you can find. These are the plastic adaptors that fit onto the end of the servo motor's output shaft:

2. Hold the corkboard in portrait orientation, and glue one servo to each side at the top, with the output shaft facing toward the front:

3. With a knife, carefully cut a notch in the end of the chopsticks:

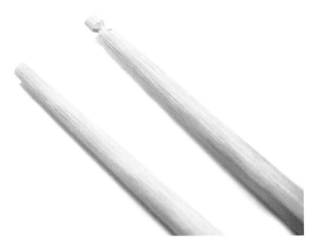

This will help to keep the string on later on!

4. Now we'll assemble the pencil head. Take your pencil and chop it down until it is very short:

5. Hot-glue the pencil to one side of the front of the servo plates:

6. Now cut your square of card so that it is just large enough to hold the servo motor, while having two holes at either side at the top to hold the string.

7. Push the servo plate onto the servo motor and then hot-glue the whole thing onto the card so the pencil is pointing away from the card:

8. Take the length of string, cut it in two, and tie one piece through each of the two holes in the card:

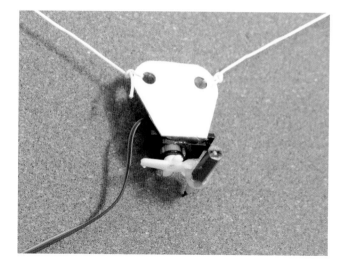

9. Now push the servo plates onto each servo, and carefully move the servo motors from one end of their travel to the other, and then bring them back and leave them in the middle position. Take the plates off and reposition them so they are entirely horizontal relative to the corkboard, then screw them on with the small screw that came with the servo motor.

10. Take the two chopsticks and place them so their ends touch on the middle of the corkboard, and mark where they meet the edge:

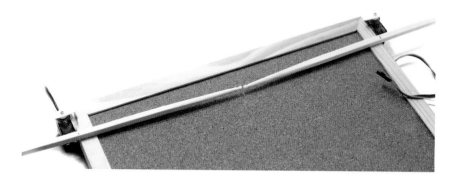

11. Turn the chopsticks around 180 degrees and place them so the marking is in the middle of each servo motor, and then tape them to each side of the double-ended servo plate that came with your servo motor:

12. Finally, tie the two bits of string for the pen assembly onto the end of the chop-sticks. You should arrange the lengths such that the pen assembly sits in the mid-dle of the corkboard:

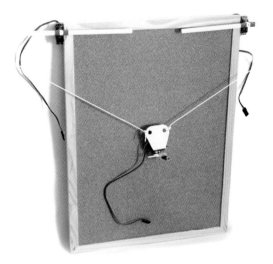

13. Pin a sheet of paper onto the bottom half of the corkboard. Put the pins halfway down the sheet to avoid the pieces of string.

14. Stand the corkboard at a 45-degree angle. You'll need gravity to help pull the pencil downward and onto the sheet of paper!

Now it's time to wire your plotter up!

1. Attach the extension wire to the servo with the pencil on it.

2. Wire the nine patch leads, three to each servo. Choose colors that are memorable: dark for ground, reds for power, and other colors for signals. See "Experiment 11: Try Out a Servo Motor" for wiring instructions on the servo.

3. Plug the Pico into the breadboard with the connector sticking out one end. This time put it as high in the breadboard as possible so you have as many sockets usable underneath it as you can get.

4. If you have a capacitor, put it between the GND and VCC pins on the Pico (GND is leftmost on the bottom of the board, VCC is the one right next to it). The capacitor will help to smooth out any bursts of current that are drawn by the servo motors.

5. Now wire all the ground leads to the GND pin near the capacitor, and all the power leads to the VCC pin.

6. Wire the left servo to pin B3, the right servo to pin B4, and the pen servo to B5. The circuit should look like this:

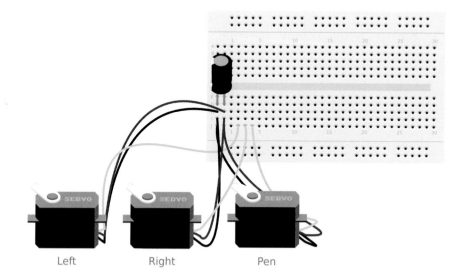

Left   Right   Pen

Software

1. Now that you're all wired up, plug the Espruino board into your PC, connect the Web IDE, enter the following on the righthand side of the IDE, and click the *Upload* button:

```
var pos = [0,0,0];
function updateServos() {
  digitalPulse(B3, 1, E.clip(1.5+motors[0], 1, 2)); // Left
  digitalPulse(B3, 1, 0); // wait for pulse
  digitalPulse(B4, 1, E.clip(1.5-motors[1], 1, 2)); // Right
  digitalPulse(B4, 1, 0); // wait for pulse
  digitalPulse(B5, 1, E.clip(1.5+motors[2], 1, 2)); // Pen
}
setInterval(updateServos, 20);
```

   All the servos should now spring into life and go to their middle positions.

   Even though we tried to set the servo motors up to their center position, we won't have it absolutely correct. Let's try to calibrate this in software:

2. Enter `motors[0]=0.1` on the lefthand side of the IDE and press `Enter`.

   The lefthand servo motor should move. Experiment with different values until you get it entirely horizontal.

3. Do the same with `motors[1]=0.1` for the righthand servo until you get that horizontal too.

4. Now try `motors[2] = 0.5;` to move the pencil. Experiment with other values until you find a value that holds the pencil nicely against the paper, and one that holds it well away from the paper.

5. Now it's time to modify our code. Enter the following code, with the values you worked out. If you've forgotten, just type `motors` on the lefthand side and Espruino will print the three values in the array:

```
var PEN_DOWN = PEN_DOWN_VALUE_HERE; // motors[2] when pen touches paper
var PEN_UP = PEN_UP_VALUE_HERE; // motors[2] when pen is away from paper
var OFFSET_LEFT = 0.1; // offset to make left servo horizontal
var OFFSET_RIGHT = -0.1; // offset to make right servo horizontal

var motors = [0,0,PEN_UP];

function updateServos() {
  getNewPosition();
  digitalPulse(B3, 1, E.clip(1.5+(motors[0]+OFFSET_LEFT), 1, 2));
  digitalPulse(B3, 1, 0); // wait for pulse
  digitalPulse(B4, 1, E.clip(1.5-(motors[1]+OFFSET_RIGHT), 1, 2));
  digitalPulse(B4, 1, 0); // wait for pulse
  digitalPulse(B5, 1, E.clip(1.5+motors[2], 1, 2)); // Pen
}
```

```
function getNewPosition() {
}

setInterval(updateServos, 20);
```

6. Upload this code and the servo motors should now jump to their correct positions.

7. We've also added a function called `getNewPosition`. We'll now use this so that we can draw some shapes. Modify `getNewPosition` as follows, and then click *Upload*:

```
var pos = 0;
var size = 0.1;

function getNewPosition() {
    // increment pos slowly between 0 and 1
    pos += 0.002;
    if (pos > 1) pos = 0;
    // Work out an angle between 0 and 360 degrees, but in radians
    var angle = pos * Math.PI * 2;
    // Now use sin and cos to move the servos in a circular motion
    motors[0] = Math.sin(angle)*size;
    motors[1] = Math.cos(angle)*size;
}
```

The pen should now be moving roughly in a circle, but it's not touching the paper.

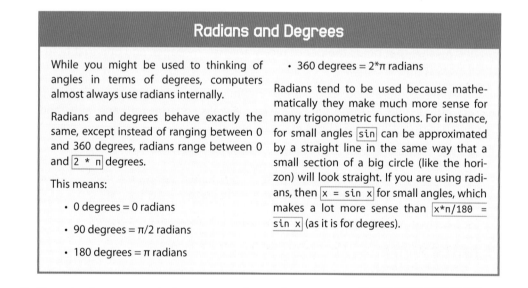

Radians and Degrees

While you might be used to thinking of angles in terms of degrees, computers almost always use radians internally.

Radians and degrees behave exactly the same, except instead of ranging between 0 and 360 degrees, radians range between 0 and `2 * n` degrees.

This means:

- 0 degrees = 0 radians

- 90 degrees = π/2 radians

- 180 degrees = π radians

- 360 degrees = 2*π radians

Radians tend to be used because mathematically they make much more sense for many trigonometric functions. For instance, for small angles `sin` can be approximated by a straight line in the same way that a small section of a big circle (like the horizon) will look straight. If you are using radians, then `x = sin x` for small angles, which makes a lot more sense than `x*n/180 = sin x` (as it is for degrees).

8. To make the pen touch the paper and start drawing, type `motors[2]=PEN_DOWN`.

It'll now touch the paper and will start to draw a circle. However, it's unlikely to be much of a circle right now. We'll fix that in a bit!

If it's not making much of a mark, you might want to experiment by sticking some pennies to the back of the pencil to give it some weight, or possibly using a softer pencil or even a pen.

9. Change `var motors = [0,0,PEN_UP];` to `var motors = [0,0,PEN_DOWN];` in your code, so the next time you upload, the pen will stay down.

10. As we created a variable called `size`, we can change the size of the circle while we're drawing it. Type `size=0.2` and the pen will suddenly jump to a new location but will then start to draw another circle of a different radius.

11. Let's experiment by drawing a square. Rather than use a formula for this, we'll use `if` statements to execute a different bit of code for each of the four edges of the square. You can just type this into the lefthand side of the IDE, and the function (and shape!) will update automatically:

```
function getNewPosition() {
    // increment pos slowly between 0 and 1
    pos += 0.002;
    if (pos > 1) pos = 0;
    // Multiply by 4, for each edge of the square
    var sq = pos*4;
    // Now set the position for each edge:
    if (sq<1) { // top edge
        motors[0] = (sq-0.5)*2*size;
        motors[1] = -size;
    } else if (sq<2) { // right edge
        motors[0] = size;
        motors[1] = (sq-1.5)*2*size;
    } else if (sq<3) { // bottom edge
        motors[0] = (2.5-sq)*2*size;
        motors[1] = size;
    } else { // left edge
        motors[0] = -size;
        motors[1] = (3.5-sq)*2*size;
    }
}
```

This is moving the left servo motor, then the right, and so on. However, it's not making a square—in fact, it's more of a wonky diamond!

12. Now experiment with different sizes, by typing `size=0.05`, `size=0.3`, etc.

You'll probably find you end up with something like this:

While for small movements we got a diamond, for larger movements there are more noticeable curved edges, and as it gets even larger things go seriously wrong and everything gets completely out of shape.

Why is this?

- We've made a plotter very quickly, and sometimes the pencil sticks on the paper and then moves quickly. As the shape gets bigger, so does the perimeter. As we're trying to draw it in the same time period we're trying to move the pencil too fast.

- If you look at the chopsticks on the servo motor as you draw a larger shape, you'll see that as the angle between them and the string gets smaller; every degree's rotation of the motor has less effect on the position of the pen. If we had been able to use a large pulley, we could have avoided this.

- The servo motors themselves are not entirely linear. They're designed for model airplanes and cars, and not for accurate movements.

If everything was taut and the servos were accurate, we could measure everything and could use a bit of geometry and math to draw perfect squares and shapes. Unfortunately, our plotter is never going to be that good, but we can improve things a lot with a very simple bit of code.

What is most obviously wrong with our squares? They're diamonds. To move upward, we need to move both servos upward at once, but at the moment we're only moving one. Similarly, to move sideways, we need to move one servo one way, and one the other.

Let's just tweak our code so it is slightly more readable, by adding l for the left servo and r for the right one:

```
function updateServos() {
  getNewPosition();
  var x = motors[0];
  var y = motors[1];
  var l = x;
  var r = y;
  digitalPulse(B3, 1, E.clip(1.5+(l+OFFSET_LEFT), 1, 2)); // Left
  digitalPulse(B3, 1, 0); // wait for pulse
  digitalPulse(B4, 1, E.clip(1.5-(r+OFFSET_RIGHT), 1, 2)); // Right
  digitalPulse(B4, 1, 0); // wait for pulse
  digitalPulse(B5, 1, E.clip(1.5+motors[2], 1, 2)); // Pen
}
```

Now, it's pretty obvious that when we draw our square's top edge by changing just motors[0], we're only moving the left servo. If we changed that so we moved both, we could make the edges of the square much straighter.

1. Replace updateServos() with the following code and re-upload it:

```
function updateServos() {
  getNewPosition();
  var x = motors[0];
  var y = motors[1];
  var l = y + x;
  var r = y - x;
  digitalPulse(B3, 1, E.clip(1.5+(l+OFFSET_LEFT), 1, 2)); // Left
  digitalPulse(B3, 1, 0); // wait for pulse
  digitalPulse(B4, 1, E.clip(1.5-(r+OFFSET_RIGHT), 1, 2)); // Right
  digitalPulse(B4, 1, 0); // wait for pulse
  digitalPulse(B5, 1, E.clip(1.5+motors[2], 1, 2)); // Pen
}
```

Now we're getting something a lot better. While it's not completely straight, the one thing that stands out is our squares are still rectangles:

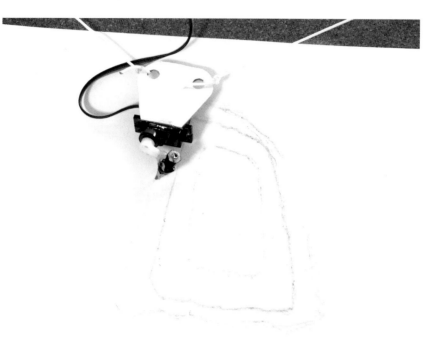

On the plotter that I made, the *square* is 3.5cm wide by 6cm high. So to fix this, we could just make sure we move less in the y axis, by multiplying it by 3.5 and dividing it by 6:

2. Change updateServos() to the following:

```
function updateServos() {
  getNewPosition();
  var x = motors[0];
  var y = motors[1];
  y = y * 3.5 / 6;
  var l = y + x;
  var r = y - x;
  digitalPulse(B3, 1, E.clip(1.5+(l+OFFSET_LEFT), 1, 2)); // Left
  digitalPulse(B3, 1, 0); // wait for pulse
  digitalPulse(B4, 1, E.clip(1.5-(r+OFFSET_RIGHT), 1, 2)); // Right
  digitalPulse(B4, 1, 0); // wait for pulse
  digitalPulse(B5, 1, E.clip(1.5+motors[2], 1, 2)); // Pen
}
```

Yours will almost certainly require different values, but after some fiddling, you should get something that looks significantly more square:

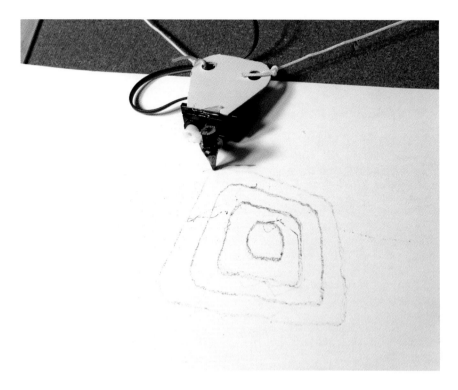

So why is the resolution in the x-axis different from the y-axis? While our servos are cheap and won't be very linear, the main reason is that in the plotter we've assembled, the chopsticks aren't angled perpendicular to the string when the pen is at the midpoint.

The math gets tricky here but just by looking at the plotter you can see that the chopsticks are horizontal when the pen is at the midpoint. This means that any rotation of the chopsticks will result in a lot more vertical movement than horizontal.

Now, with some simple trial and error, you've got a plotter that will plot different shapes. At the moment it's really not that exciting, as it only plots the same shape until you upload more code.

Instead of making just one `getNewPosition` function, let's make a function called `move` that sets `getNewPosition` to some code that will move the pencil to a certain location, and will call a callback function when it's done.

3. Replace `getNewPosition` with the following and upload again:

```
function doNothing() {
  // do nothing
}
```

```
var getNewPosition = doNothing;

function move(x, y, callback) {
  // First, get the old positions
  var oldx = motors[0];
  var oldy = motors[1];
  // Now work out the distance using Pythagoras
  var dx = oldx-x;
  var dy = oldy-y;
  var d = Math.sqrt(dx*dx + dy*dy);
  // Make sure we move at the right speed, not
  // too fast or slow!
  var speed = 0.002/d;
  // and now have 'pos', our position in the line
  var pos = 0;
  // Finally, set the getNewPosition function to something
  // that will draw a line

  getNewPosition = function() {
    pos += speed;
    if (pos>1) {
      // If we've finished, stop and
      // call the callback
      pos = 1;
      getNewPosition = doNothing;
      if (callback) callback();
    }
    // Set the motor positions up by interpolating
    // between oldx and x, oldy and y
    motors[0] = oldx*(1-pos) + x*pos;
    motors[1] = oldy*(1-pos) + y*pos;
  };
}
```

Now, nothing will happen.

4. However, if you type `move(0.2, 0)` on the lefthand side of the IDE, the pen will now move to the right.

5. Type `move(0, 0.2)`, and the pen will move diagonally down.

6. Type `move(-0.2, 0)`, and you should now have a `V`.

7. Type `move(0.2, 0)` again to go back to the beginning and draw a triangle.

However, our function has a callback, so we can chain calls together. For instance, the following will draw the triangle and print `Done!` when it is finished:

```
move(0.2, 0, function() {
  move(0, 0.2, function() {
    move(-0.2, 0, function() {
      move(0.2, 0, function() {
        console.log("Done!");
```

```
          });
        });
      });
    });
```

8. We can now rewrite our square-drawing function like this:

```
function penDown(yes) {
  if (yes) motors[2] = PEN_DOWN;
  else motors[2] = PEN_UP;
}

function square(x,y,size, callback) {
  move(x-size, y-size, function() {
    penDown(true);
    move(x+size, y-size, function() {
      move(x+size, y+size, function() {
        move(x-size, y+size, function() {
          move(x-size, y-size, function() {
            penDown(false);
            if (callback) callback();
          });
        });
      });
    });
  });
}
```

9. If you now type `square(0, 0, 0.1)`, you'll draw a square in the same place as the original. The pen will go down at the start and up at the end.

10. You can put a new square next to it with `square(0.2, 0, 0.1)`, or one underneath with `square(0, 0.2, 0.1)`.

    In fact, now that you have a working plotter, you could even use it to draw text. For instance, `A` would be:

```
function drawA(callback) {
  move(-0.1, 0.1, function() {
    penDown(true);
    move(0, -0.1, function() {
      move(0.1, 0.1, function() {
        penDown(false);
        move(-0.05, 0, function() {
          penDown(true);
          move(0.05, 0, function() {
            penDown(false);
            if (callback) callback();
          });
        });
      });
    });
  });
```

```
    });
  }
```

What else can you draw? You could change the funcion to draw stick figures, faces, or could even add one function per letter of the alphabet.

If you have one function per character, you could write text by chaining the functions together in the right other. For instance, to draw ABC you could do:

```
drawA(function() {
  drawB(function() {
    drawC(function() {
    });
  });
});
```

After you've read Chapter 15 (connecting with WiFi), you could even go back and connect your plotter to the internet!

This is an amazingly simple plotter, and it has some big problems:

Nonlinearity

Using chopsticks means that the angle between the string and each chopstick changes when the chopsticks move, and so a 1-degree movement at one end of travel will move the string a different amount compared to a 1-degree movement in the middle. While you could try to account for this change with math, it would be a lot easier to use a pulley, which would always move the string the same amount, no matter where it was!

Accuracy

Cheap servo motors aren't very accurate. They are unlikely to be able to position the chopstick accurately to less than 1 degree. To make a much more useful plotter, you could use stepper motors instead. Stepper motors can rotate multiple times, so in combination with a small pulley they could provide much more accuracy.

Digital Pinhole Camera

10

At the heart of pretty much all current digital cameras is a piece of silicon with millions of distinct light sensors on it. The camera lens focuses light onto the silicon where each sensor detects the amount of light it receives, and the color of that light. The information from those millions of sensors is then read into a small computer in the camera, which creates an image.

While this is probably the best way of making a digital camera, it isn't the only way. If you are trying to take a picture of a scene that isn't moving, you can use a single light sensor that you move to each location in the image.

That's the kind of camera that we're going to make here. We'll skip the lens and will instead use a small hole at the end of a long tube, allowing the sensor to view only a small area of the scene at once. We're basically making a one-pixel pinhole camera!

Experiment 15: Making a Digital Camera

You'll need:

- A cheap, plastic ballpoint pen
- Black masking tape
- An elastic band
- A block of wood, roughly 7cm x 7cm x 7cm
- A light-dependent resistor (LDR), also known as a photoresistor
- 1 meter of solid core wire
- A 100k Ohm resistor
- 2x servo motors (not continuous rotation)

- Breadboard
- 7 jumper leads
- An Espruino Pico
- A 47uF 6v (or higher) capacitor (optional)

Follow these steps to assemble the wand:

1. Cut the solid core wire into two equal lengths, and strip 10mm off one end of each wire:

2. Fold the stripped end of each wire at 90 degrees, and twist it onto each of the LDR's wires. If you're happy soldering, you'll probably want to make a solder joint instead:

3. Now wrap the masking tape twice around one leg of the LDR to insulate it, and then wrap it around both legs:

4. Take your cheap plastic ballpoint pen, and pull the front off of it:

If possible, drill roughly a 2mm-diameter hole in the back end of the pen. If not, just pull the back out. Be careful: the larger the hole the more blurry your picture will be!

5. Take your LDR and push it into the open end of the pen's body, then use masking tape to hold it in:

Wrap the masking tape all the way to the back to ensure that no light will be able to get to the LDR, except by the hole in the end:

You should now have a wand assembly that will be able to measure the light that is coming from the direction in which it is pointed.

6. Strip 6mm off the other end of the wires—enough that they'll fit snugly into a bit of breadboard.

7. Use a two-ended plate from your servo motor (like you used in Chapter 9) and place it on the wand, roughly 1/3 of the way away from the sensor. Wrap around each end with masking tape. Your wand should now look like this:

Follow these steps to assemble the camera:

1. Now you just need to make a base that'll move the wand around. Take your two servos, and place them together with their output shafts at right angles. Now use an elastic band (or the masking tape) to wrap them together:

2. Use a second servo plate and two screws that came with the servos to secure the plate to your wooden block at a corner near the top:

3. Now position the servos and wand, so the servo attached to the block of wood rocks the wand up and down, and the servo attached to the wand moves it side to side:

4. Finally, place the whole arrangement pointing at a your test scene. Since we're not making a very good camera, you'll need to use a very simple test picture with bold black outlines on white:

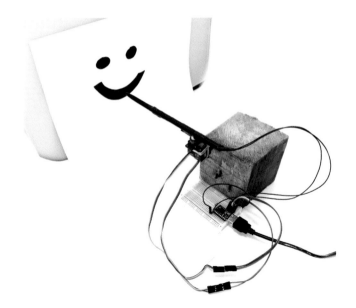

Follow these steps to wire your pinhole camera:

1. Plug the Pico into the breadboard with the connector sticking out one end. This time put it as high in the breadboard as possible so you have as many sockets usable underneath it as you can get:

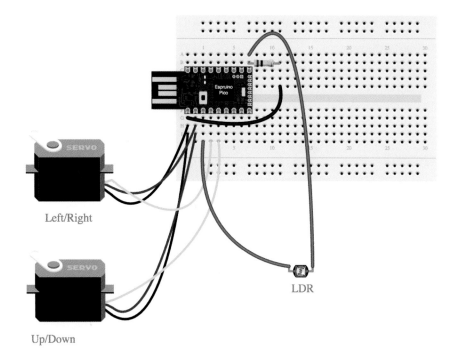

2. If you have a capacitor, put it between GND and VCC pins on the Pico (GND is left-most on the bottom of the board, VCC is the one right next to it). This will help to smooth out any surges of current drawn by the servo motors.

3. Now wire all the ground leads to the GND pin near the capacitor, and all the power leads to the VCC pin.

4. Wire the left/right servo to pin B3 and the up/down servo to pin B4.

5. Now use the remaining jumper lead to connect from the Pico's GND to an upper column of the breadboard just to the right of the Pico.

6. Connect the 100k Ohm resistor between this column and the one to the left of it (pin A5 on the Pico).

   We used a 10k for the robot, but this time we're using a 100k resistor because the amount of light we expect our sensor to get right at the bottom of the black tube is really small.

7. Connect the two wires from the wand, one to 3.3v on the Pico (third from the left at the bottom), and one to pin A5.

Now you're ready to go!

Follow these steps to prepare the software:

1. Add the following code on the righthand side of the IDE, and click *Upload*:

```
function updateServos() {
  var x = 0;
  var y = 0;
  digitalPulse(B3, 1, E.clip(1.5+x, 1, 2));
  digitalPulse(B3, 1, 0); // wait for first pulse to finish
  digitalPulse(B4, 1, E.clip(1.5+y, 1, 2));

}
setInterval(updateServos, 20);
```

This will simply center both your servos.

2. If your wand is now not pointing straight ahead, pull the servos from the wand and block of wood and reposition them so that the wand is pointing straight ahead.

Now we can test the light sensor. As we did with the robot, we'll use `analogRead` to read the analog value.

3. Type `analogRead(A5)`, and you'll get a reading. If you put something white in front of the sensor, you should get a much higher reading than if you put something black.

Unlike with the robot, this time we put the resistor and the light sensor in the opposite order (the resistor pulling down, with the light sensor pulling up). Now, this causes the voltage (and so the value from `analogRead`) to rise when there is more light, and fall when there is less.

Our next step is to try to scan the wand left and right, so that we'll be able to read in what it sees when pointing in different directions.

4. Modify the code on the righthand side of the IDE to the following:

```
// How detailed our picture will be
var WIDTH = 48;
var HEIGHT = 48;
var PIXELS = WIDTH*HEIGHT;

// The position in our scan
var px=0, py=0;

function readPixel() {
  // not doing anything yet
}

function updateServos() {
  readPixel();

  /* Bring px and py into the
  right range for the servo motors */
```

```
        var x = ((px/WIDTH) - 0.5) / 3;
        var y = ((py/HEIGHT) - 0.5) / 3;

        // And move the servos
        digitalPulse(B3, 1, E.clip(1.5+x, 1, 2));
        digitalPulse(B3, 1, 0);
        digitalPulse(B4, 1, E.clip(1.5+y, 1, 2));

        /* Move to the next position. Go right */
        px++;
        // or if we're at the end of the line,
        // go back to the start
        if (px>=WIDTH) {
          px=0;
          py++;
        }
        if (py>=HEIGHT) {
          /* If we got to the end, don't do anything
             else. Stop calling updateServos */
          clearInterval(scanInterval);
        }
      }

    var scanInterval = setInterval(updateServos, 20);
```

When you click *Upload* now, the wand will sweep slowly from left to right and will then quickly return to the left where it'll start again. When the wand finally gets to the end it'll stop.

While the wand is moving, you can check the values of px and py to see what pixel Espruino thinks it's scanning. However, at the moment we're not reading any information from the wand.

The next step is to read each pixel as the wand points at it, and to store that pixel.

5. To do this this, modify readPixel by adding the following:

```
    // Our pixel data
    var data = new Float32Array(PIXELS);

    function readPixel() {
      var light = analogRead(A5);
      // work out where in the array it should go
      var idx = px + (py*WIDTH);
      // save the data away
      data[idx] = light;
    }
```

Float32Array allows us to store the pixel data efficiently in memory. If we used a normal array, we wouldn't be able to fit the whole picture in Espruino's memory at once!

When you upload now, you won't immediately see anything different, but you can type `data` to show all the pixels that have been read in as the wand was scanned. Any pixels not scanned will be set to `0`.

By itself, this isn't very useful; all we have are numbers! What we want to do is reconstruct the data as an image. Since the values we are reading in don't fill the full range of our A to D converter, first we'll need to adjust them so that they range from completely black to completely white so we can see them properly.

To display something in the terminal window is a bit tricky as we don't have any graphical output and can only output characters. The easiest way to get something visible is to rank some characters by their apparent brightness, for instance, `.:;*@#`—then when they are printed on the screen you'll just about be able to make out a picture!

6. Add the following to the end of your program, and upload again:

```
// Draw our pixels out to the screen
function draw() {
    /* We have to use characters to represent
    each shade of color, so we're putting some
    characters in a string that get progressively
    more 'dense' */
    var shades = " .:;*@#";
    /* Work out the maximum and minimum
    values, so we can scale the image
    brightness properly */
    var min = data[0];
    var max = data[0];
    data.forEach(function(pixel) {
        if (pixel < min) min = pixel;
        if (pixel > max) max = pixel;
    });
    // Now we can print the data out, line by line
    var n = 0;
    for (var y=0;y<HEIGHT;y++) {
        var str = "";
        for (var x=0;x<WIDTH;x++) {
            var light = (data[n]-min)/(max-min);
            var shade = Math.floor(light*shades.length);
            str += shades[shade];
            n++;
        }
        console.log(str);
    }
}
```

7. Now, you can type `draw()` and a very low-res version of the picture from the camera will be displayed:

Hopefully you can now see the outline of the picture you have scanned, but it's not very easy to see. To display a proper image, you can dump the raw data and then use a web browser on your PC to view it.

8. Add the following code, and then run getData():

```
function getData() {
  var min = data[0];
  var max = data[0];
  data.forEach(function(pixel) {
    if (pixel < min) min = pixel;
    if (pixel > max) max = pixel;
  });
  // Print the data out, line by line
  var n = 0;
  for (var y=0;y<HEIGHT;y++) {
    var str = "";
    for (var x=0;x<WIDTH;x++) {
      var light = Math.round((data[n]-min)*255/(max-min));
      str += light+",";
      n++;
    }
    console.log("["+str+"],");
  }
}
```

You'll now get a big list of data. It's been rescaled to a set of numbers between 0 and 255.

9. Now open a web browser pointing to *http://jsfiddle.net*.

10. Type `<canvas width="512" height="512"/>` into the box titled HTML .

11. Add the following text into the box titled JAVASCRIPT on the website:

```
var data = [
// pixel data
];

var ctx = document.getElementsByTagName("canvas")[0].getContext("2d");
data.forEach(function(row,y) {
 row.forEach(function(pixel,x) {
   ctx.fillStyle = "rgb("+pixel+","+pixel+","+pixel+")";
   ctx.fillRect(x*8, y*8, 8, 8);
 });
});
```

12. Now highlight the numbers that getData returned in the Web IDE (including the square brackets), and paste them into the code that you entered in jsfiddle where the comment // pixel data was.

13. Finally, click Run in the top left. You should now see your image in proper grayscale:

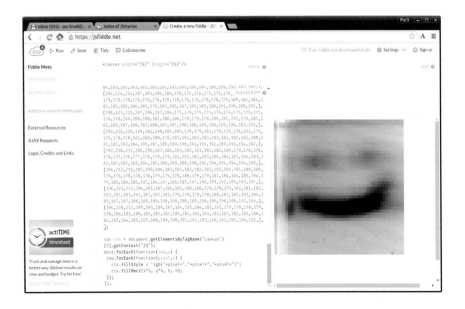

You may have noticed that the code we entered didn't leave any time for the wand to move back from right to left after each newline, and you can now see the effects of that in the preceding image. On the far left, there's a very narrow, backwards version of the image, caused by the wand moving quickly from right to left over the image in order to get back to the start of the newline.

There are two easy ways of fixing this: you could wait until the wand had moved back to the correct position, or you could make the wand move in a zig-zag, scanning from left to right and then from right to left. The problem with the second option is that the servo motors don't reach the correct position immediately, causing the wand to lag behind the position that is in our variables. If we were to scan in two directions then rather than having the same amount of lag for each line, alternate lines would be different, causing the image data not to line up:

Complete Listing

```
// How detailed our picture will be
var WIDTH = 48;
var HEIGHT = 48;
var PIXELS = WIDTH*HEIGHT;

// The position in our scan
var px=0, py=0;

// Our pixel data
var data = new Float32Array(PIXELS);

function readPixel() {
  var light = analogRead(A5);
  // work out where in the array it should go
  var idx = px + (py*WIDTH);
  // save the data away
  data[idx] = light;
}
```

```
function updateServos() {
  readPixel();

  /* Bring px and py into the
  right range for the servo motors */
  var x = ((px/WIDTH) - 0.5) / 3;
  var y = ((py/HEIGHT) - 0.5) / 3;

  // And move the servos
  digitalPulse(B3, 1, E.clip(1.5+x, 1, 2));
  digitalPulse(B3, 1, 0);
  digitalPulse(B4, 1, E.clip(1.5+y, 1, 2));

  /* Move to the next position. Go right */
  px++;
  // or if we're at the end of the line,
  // go back to the start
  if (px>=WIDTH) {
    px=0;
    py++;
  }
  if (py>=HEIGHT) {
    /* If we got to the end, don't do anything
       else. stop calling updateServos */
    clearInterval(scanInterval);
  }
}

var scanInterval = setInterval(updateServos, 20);

// Draw our pixels out to the screen
function draw() {
  /* We have to use characters to represent
  each shade of color, so we're putting some
  characters in a string that get progressively
  more 'dense' */
  var shades = " .:;*@#";
  /* Work out the maximum and minimum
  values, so we can scale the image
  brightness properly */
  var min = data[0];
  var max = data[0];
  data.forEach(function(pixel) {
    if (pixel < min) min = pixel;
    if (pixel > max) max = pixel;
  });
  // Now we can print the data out, line by line
  var n = 0;
  for (var y=0;y<HEIGHT;y++) {
    var str = "";
    for (var x=0;x<WIDTH;x++) {
      var light = (data[n]-min)/(max-min);
      var shade = Math.floor(light*shades.length);
      str += shades[shade];
```

```
        n++;
      }
      console.log(str);
    }
  }

  function getData() {
    var min = data[0];
    var max = data[0];
    data.forEach(function(pixel) {
      if (pixel < min) min = pixel;
      if (pixel > max) max = pixel;
    });
    // Print the data out, line by line
    var n = 0;
    for (var y=0;y<HEIGHT;y++) {
      var str = "";
      for (var x=0;x<WIDTH;x++) {
        var light = Math.round((data[n]-min)*255/(max-min));
        str += light+",";
        n++;
      }
      console.log("["+str+"],");
    }
  }
```

So what can you do now? You could experiment with different designs of wand to try to get a sharper picture. For example, you could use a smaller hole in the end of the wand (perhaps an actual pinhole), and could use a tube that was matte black on the inside (as the shiny pen will reflect light coming from other angles).

You could also add the same light sensor to the plotter from Chapter 9, and could try to scan documents with it! Some early scanners did exactly that (moving a single sensor over the entire page), although modern scanners tend to use a line of thousands of sensors (called a linear CCD) that they move down the page, scanning each row of pixels in turn.

Printer

11

In the first part of this section we made a simple plotter, where we could draw lines by moving a pen anywhere in two dimensions. However, the printers you use every day don't do this. They *scan* a print head down the page. Much as we did with the camera, they move from side to side and then down the page, covering the area but only applying ink where needed.

Making your own printer can be quite easy. Here, we're going to use a screw thread to let us print on some paper with just a motor, position sensor, and one actuator to move the pen up or down.

Because we've used servo motors for previous projects, we'll use them here. However, you could use any geared motor, and even a solenoid to control whether the pen touches the paper or not.

Experiment 16: Making a Printer

You'll need:

- 35cm, threaded rod, 1 nut, and 1 penny washer to fit it (M6, M8 or similar is perfect)
- A sturdy cardboard box, 30cm on one side
- Hot glue
- A Pringles can or whisky tube
- A *9g* size, continuous rotation servo motor
- A *9g* size RC servo motor (*not* continuous rotation)
- A felt-tip pen

- A light-dependent resistor (LDR)
- A 10k Ohm resistor
- Breadboard
- An Espruino Pico
- 6 patch wires

Follow these steps to assemble the printer:

1. Cut the top off the cardboard box, and then cut the box in half so you have a 30cm-long shelf about 1.5x as high and deep as the tube is in diameter. Strengthen it if needed.

2. Cut the cardboard tube so that it is 13cm long. The tube should fit inside the box twice, with a bit of room left over.

3. Drill a hole in the center of each end of the tube, just large enough to put the threaded rod through.

4. Put the threaded rod through the tube, with a nut partway up. If you have a second nut, place it on the rod after the tube.

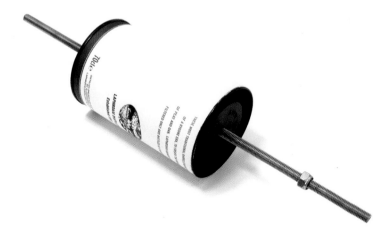

5. Take a small servo plate for the continuous rotation servo, cut the long edges off, and hot-glue it onto the end of the threaded rod as centrally as you can.

6. Cut a large hole in the left and right sides of the cardboard, high enough that the rod can fit through with the tube on it, leaving around 2cm underneath. The hole should be a few millimeters larger than the rod.

7. On the right-hand side of the box, hot-glue the penny washer over the hole. This will be your righthand bearing:

8. On the lefthand side, extend the hole into an offset rectangle so that you can insert and hot glue the continuous rotation servo motor with its axle where the hole was:

9. Now you need to make a *sled* that the pen will be mounted on. Cut a thin, straight strip of card and hot-glue it to the very edge of the box at the bottom. This will hold the sled in place.

10. Now to make the sled. Fold a piece of card into an L-shape that fits inside the box next to the strip of card (with spare room on the left). Cut and hot-glue two triangular bits of card to strengthen it.

11. Cut another thin, straight strip of card and hot-glue it to the box on the other side of the sled, making sure you can still slide the sled easily to the left and right. This will stop the sled from twisting from side to side.

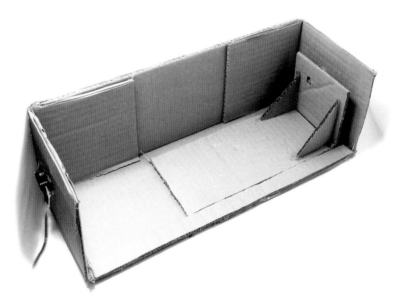

12. Now place the sled at the end of the box, mark through the hole in your box (around the washer) with a pen, and cut out the hole.

13. Assemble everything as shown:

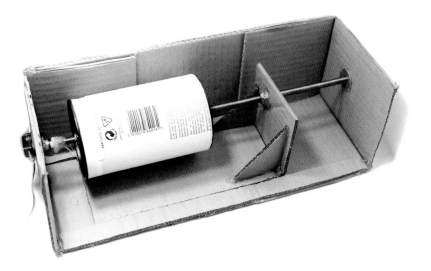

Hot-glue the nut onto the sled. This will make the sled move slightly as the drum with the paper on it rotates.

 The nut shown is a nyloc nut, but the locking part has been removed. Ideally you should use a nonlocking nut.

14. Now it's time to fit the pen to the sled. Take a felt-tip pen and hot-glue it to a servo plate from the noncontinuous rotation servo, as close to the end as possible.

You may even be able to cut the cap of pen so that it can still fit around the servo plate.

15. Cut a strip of cardboard and fold it into a triangular prism shape as wide and high as the servo motor. This will hold the motor at a right angle to the sled:

16. Now fit the pen onto the servo motor and glue the servo motor to the cardboard triangle with the axle sticking out of the side toward the top:

17. Glue the whole assembly onto the sled, positioned so that the servo motor is able to move the pen onto and away from the tube.

   We're almost there, but as we're using the continuous rotation servo motor, the Espruino will have no idea where the tube is in its rotation. We need to add a sensor.

18. Add two wires to the light-dependent resistor (LDR) up as you did for the camera in Chapter 10.

19. Cut a hole in the back edge of the cardboard and push the LDR through. Hot-glue the LDR in, so that it sits about 0.5cm in from the edge of the tube, and 0.5cm away:

20. Wrap white tape (or white stickers) around the edge of the tube where the light sensor is.

21. Color a big square of the white tape black. This will be detected by the light sensor so that the printer is able to detect when the tube is rotated by 360 degrees.

22. Now wire the servos and LDR up exactly as you did for the camera in Chapter 10, except this time wire the continuous rotation servo (for the paper) to B3, and the normal servo (for the pen) to B4:

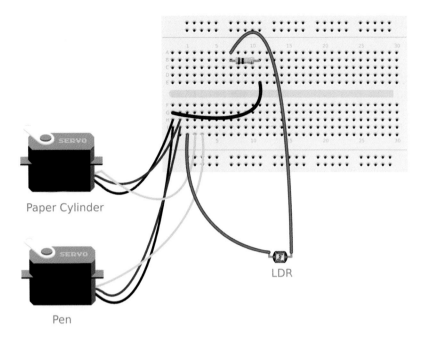

Paper Cylinder

LDR

Pen

Software

This time we're going to control the servo motors a bit differently. We'll use the Espruino board's built-in PWM hardware with the `analogWrite` command. This will allow us to let the hardware send pulses while the software is doing other things.

1. First, try `analogWrite(B3, 1.5/20, {freq:50})`. This shouldn't have any effect at all, but if the tube starts moving then tweak the potentiometer on the back of it to stop it (as you did for the robot in Chapter 8).

2. Now, try either `analogWrite(B3, 1/20, {freq:50})` or `analogWrite(B3, 2/20, {freq:50})` and find out which one makes the tube move so that the pen carriage moves away from the end with the light sensor on it. We'll call this *forward*.

3. Create three functions on the righthand side of the IDE as follows (and swap the numbers if needed to ensure that the tube rotates the correct way):

```
function forward() {
  analogWrite(B3, 1/20, {freq:50});
}

function back() {
  analogWrite(B3, 2/20, {freq:50});
}

function stop() {
```

```
    digitalWrite(B3, 0);
  }
```

4. The next step is to set up the pen. Remove it from the servo motor and type `ana` `logWrite(B4, 1.5/20, {freq:50})`. This will move the motor to its mid position. Now re-add the pen so it's not pressing onto the paper, and experiment with values other than `1.5` (remember to stick with numbers between 1 and 2) to find two good values for when the pen is on the paper (pen down) and when it's not (pen up).

5. Create functions using your two values:

```
function penUp() {
  analogWrite(B4, 1.45/20, {freq:50});
}

function penDown() {
  analogWrite(B4, 1.55/20, {freq:50});
}
```

6. Now you can upload your code, and set the motor running by typing `forward()` or `stop()`.

The next step is to try to detect each time the marker we colored in moves past. To do this, we'll keep a variable called `lightAverage`, which is the average amount of light we're seeing in the light sensor. When the light that's reflected back drops suddenly, we'll assume that the black marker has passed underneath the sensor, and when it rises back up we'll assume that the marker has completely gone under.

Add the following:

```
function foundMarker() {
  // work out how long it's been since the last marker
  var t = getTime();
  var d = t - lastMarkerTime;
  lastMarkerTime = t;
  console.log("Found marker, "+d+" sec");
}

function lightChecker() {
  // work out how much light is reflected
  // from the paper
  var light = analogRead(A5);
  if (light < lightAverage-0.05) {
    // If the light is significantly less
    // than the average, we've found the marker
    if (!hasFoundMarker) {
      hasFoundMarker = true;
      foundMarker();
    }
```

```
    } else if (light > lightAverage-0.03) {
        // If it jumps back up, we've passed it
        hasFoundMarker = false;
    }
    // update the average
    lightAverage = lightAverage*0.99 + light*0.01;
}

setInterval(lightChecker, 10);
```

7. Now upload the code again and type `forward()`.

   If all goes well, a series of lines should output each time the marker passes underneath the sensor, like:

```
Found marker, 0.77991580963 sec
Found marker, 0.77999591827 sec
Found marker, 0.77989006042 sec
Found marker, 0.77994823455 sec
Found marker, 0.76992893218 sec
Found marker, 0.77994823455 sec
Found marker, 0.76997661590 sec
Found marker, 0.77989292144 sec
```

   They should all have roughly the same time period, showing that the tube is rotating at the same speed and the code for the sensor is working properly.

   If not, you might need to modify the preceding code such that the thresholds (`0.05` and `0.03`) are different. You can look at the value of `analogRead(A5)` when the marker is under the light sensor and when it has moved away to get an idea what you should be using. If the difference in value reported by `analogRead` is very small, you might need to:

   • Reposition your light sensor to point more directly at the marker

   • Make the black area of the marker larger

   Now that we've got that working, we can start piecing everything together. To start, let's just color in a long column.

8. Add the lines `var pixelInterval;` and `var x = 0, y = 0;` above `foundMarker` and then add to `foundMarker` so it looks like this:

```
function foundMarker() {
    // work out how long it's been since the last marker
    var t = getTime();
    var d = t - lastMarkerTime;
    lastMarkerTime = t;
    console.log("Found marker, "+d+" sec, line "+y);

    if (pixelInterval)
        clearTimeout(pixelInterval);
```

```
    // move on to next line
    y++;
    x=0;
    // Execute for each 'pixel'
    pixelInterval = setInterval(function() {
      if (x < 2)
        penDown();
      else
        penUp();
      x++;
    }, d*1000/10); // 10 pixels/line
  }
```

On each line, we'll set x to 0, and a timer will be started that will divide the rotation into 10 distinct pieces. Each time the timer is called, we'll add 1 to x for each piece.

At the moment we're only putting the pen down if x is below 2 (so the first 2 *divisions* only), which should cause it to draw a column.

9. Now wind the tube back to the beginning (either manually, or by typing back()).

10. When you upload and type forward() , the pen should be put down and raised once for every revolution of the tube, slowly building up a picture like this:

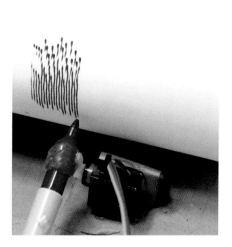

But this can easily be extended. First, we'll use Espruino's built-in graphics to give us a simple image.

11. Enter the following at the end of the righthand side of the IDE:

```
var g = Graphics.createArrayBuffer(96,48,1);
g.setFontVector(48);
g.drawString("=P",0,0);
```

However, it's difficult to make sure that we've positioned the text correctly, and since it's so difficult to position the paper on the tube we want to make sure that what we want to draw is correct!

12. Enter this function called draw. It's very similar to the code we used in the camera project earlier:

```
function draw() {
  for (var y=0;y<g.getHeight();y++) {
    var s = "|";
    for (var x=0;x<g.getWidth();x++)
      s += g.getPixel(x,y) ? "#" : " ";
    console.log(s+"|");
  }
}
```

13. Now if you upload and type **draw()** you should see something like this:

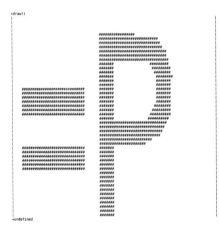

This shows that our =P text is lined up neatly within the | characters at either side, and so should fit on our paper.

14. The next step is to modify foundMarker such that it can use getPixel too, rather than just checking x < 2 as it was previously:

```
function foundMarker() {
  // work out how long it's been since the last marker
  var t = getTime();
  var d = t - lastMarkerTime;
  lastMarkerTime = t;
  console.log("Found marker, "+d+" sec, line "+y);
```

```
    if (pixelInterval)
      clearTimeout(pixelInterval);

    // move on to next line
    y++;
    x=0;
    // Execute for each 'pixel'
    pixelInterval = setInterval(function() {
// --------------------------- Only the lines below have changed
      if (g.getPixel(x,y))
        penDown();
      else
        penUp();
      x++;
    }, d*1000*0.75/g.getWidth());
// --------------------------- Only the lines above have changed
  }
```

15. If you upload again, (having returned the tube to the start again) and type `for ward()`, something different will happen this time:

    - For the first few rotations, nothing will happen (you can see this because when we drew the image to the console, the first few lines were blank).

    - At first, the pen will only go down once, as the top part of the P is printed.

    - Then the pen will go down twice, as the P opens out.

    - And then you'll get three presses, as the = and the P start to be printed.

    - And so on until the end.

    The final result should look like this:

 If the tube rotates unevenly, it will cause the pen to move up and down at the wrong points, producing the jitter you can see in the complete printed image. If this happens to you, you can change the `penDown` *function so that it presses the pen more lightly against the tube.*

It's now easy to experiment with different images, by replacing `g.drawString("=P",0,0);` with your own code. There are lots of interesting functions other than `drawString` that you can use in Espruino's graphics library (*http://www.espruino.com/Reference#Graphics*).

Some other interesting changes might be:

- Stop the rotation of the tube when printing has finished (`y` is greater than or equal to `g.getHeight()`).

- Use `back()` to move the pen back to the start, and have it stop at the correct place.

- Use the data gathered from the camera in the last experiment to print a simple image captured from the real world.

- Replace the continuous rotation servo with a stepper motor, allowing you to step more slowly (without a rotation sensor), and draw more accurately. I didn't do that in this chapter because you need to be sure that you have a powerful enough stepper motor to overcome the friction of the pen against the tube.

Complete Listing

```
function forward() {
  analogWrite(B3, 1/20, {freq:50});
}

function back() {
  analogWrite(B3, 2/20, {freq:50});
}

function stop() {
  digitalWrite(B3, 0);
}

function penUp() {
  analogWrite(B4, 1.45/20, {freq:50});
}

function penDown() {
  analogWrite(B4, 1.55/20, {freq:50});
}

var lightAverage = analogRead(A5);
var hasFoundMarker = false;
var lastMarkerTime = getTime();
var pixelInterval;
var x = 0, y = 0;

function foundMarker() {
  // work out how long it's been since the last marker
  var t = getTime();
  var d = t - lastMarkerTime;
  lastMarkerTime = t;
  console.log("Found marker, "+d+" sec, line "+y);

  if (pixelInterval)
    clearTimeout(pixelInterval);

  // move on to next line
  y++;
  x=0;
  // Execute for each 'pixel'
  pixelInterval = setInterval(function() {
    if (g.getPixel(x,y))
      penDown();
    else
      penUp();
    x++;
  }, d*1000*0.75/g.getWidth());
}

function lightChecker() {
  // work out how much light is reflected
  // from the paper
```

```
  var light = analogRead(A5);
  if (light < lightAverage-0.05) {
    // If the light is significantly less
    // than the average, we've found the marker.
    if (!hasFoundMarker) {
      hasFoundMarker = true;
      foundMarker();
    }
  } else if (light > lightAverage-0.03) {
    // If it jumps back up, we've passed it
    hasFoundMarker = false;
  }
  // update the average
  lightAverage = lightAverage*0.99 + light*0.01;
}

setInterval(lightChecker, 10);

var g = Graphics.createArrayBuffer(96,48,1);
g.setFontVector(48);
g.drawString("=P",0,0);

function draw() {
  for (var y=0;y<g.getHeight();y++) {
    var s = "|";
    for (var x=0;x<g.getWidth();x++)
      s += g.getPixel(x,y) ? "#" : " ";
    console.log(s+"|");
  }
}
```

PART IV

Communication

The internet is now a massive part of our lives, but smart things are still communicating with each other all around us without an internet connection.

And often they're doing so in a surprisingly simple way that's not all that different from the Morse code that we were using 180 years ago!

Wired Communication

12

So far, we've controlled things with an Espruino board, but apart from the USB connection to the PC we haven't actually communicated with any other devices.

Suppose we wanted to communicate with another Espruino board, turn a television on and off, or even connect to the internet. We need some way of turning our data (at the basic level, made up out of a series of bits) into something that we can transmit, and then we need a way to reconstruct it.

Perhaps the most obvious example of transmitting data is Morse code. A series of short and long beeps is used to transmit characters of text over the radio.

For example, if we use . for a short beep, and - for a long beep, the codes for various characters are:

A .-	B -...	C -.-.	D -..	E .
F ..-.	G --.	H	I ..	J .---
K -.-	L .-..	M --	N -.	O ---
P .--.	Q --.-	R .-.	S ...	T -
U ..-	V ...-	W .--	X -..-	Y -.--	Z --..

So the signal for SOS (the distress signal that you'll send if your ship is sinking) is ... --- ....

Clocking

However, this gives us an interesting problem. Some Morse code radio operators are better than others, and so are faster. How do you tell the difference between a slow operator sending an *E* character as a single short beep, and an experienced operator sending a *T* character as a long beep (but quickly)?

We'd just have to hope that a message will contain a mix of short and long beeps so that we can tell which is which, and that the operator never slowed down or sped up. Such hope might work fine for humans, but it's not good for computers.

Instead, there are two main solutions: have an agreed speed (which we usually call a *bit rate* or *baud rate*), or have a separate signal that tells us when we're sending a new bit of data (Figure 12-1). This is normally called a *clock* signal.

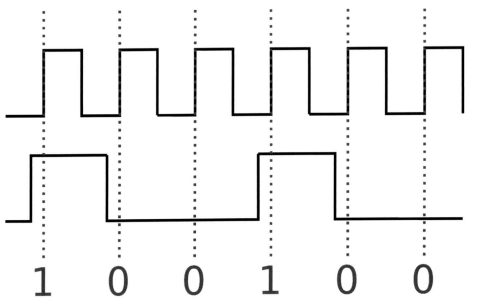

Figure 12-1 *Clocked data like SPI has a separate wire to determine when data should be read*

SPI (the Serial Peripheral Interface) is a standard that's used internally in a lot of devices. For instance, the accelerometer in your phone might use it, and even SD cards support it as a fallback mode (if the faster SDIO standard isn't supported by the device they're plugged into). SPI has three main wires (plus a ground connection). There is one for transmitting data, one for receiving data, and one for a clock. The clock tells the *slave* device when to read and write its data.

However, infrared remote controls can't have a clock signal since they're just sending one wavelength of light from an infrared LED. They also have to transmit short pulses of light at around 38kHz to help distinguish their signals from ambient noise (for example, a shadow passing over the TV's IR receiver). To get around this, they send bursts of infrared pulses that last different amounts of time (see Figure 12-2). While different infrared remote controls vary, they usually send pulses of light for less than 0.8ms to represent a 0 bit, and a longer series of pulses to represent a 1 bit.

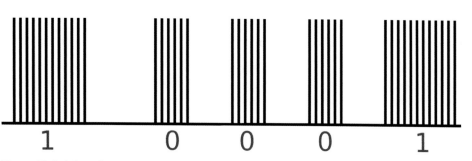

Figure 12-2 *Infrared remote control signals*

If you're going to agree on a fixed transmission speed in advance, you can be more efficient, though. Suppose we have agreed on 1000 baud (1000 bits per second). Once we see a bit of data appear, we can just check 1/1000th of a second later for the next bit, and again after that (Figure 12-3). There's no need to have pulses of varying lengths at all.

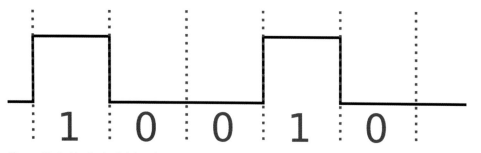

Figure 12-3 *Unclocked data, always transmitted at an agreed* bit rate

While Morse code has varying amounts of dots or dashes for different characters (which was handy for operators, as common characters tended to be shorter), on a computer we generally assume that we'll be sending compressed data (so every character we send will be equally likely to turn up). Because of this, it's much easier to just have a fixed *word length*—usually 8 bits to correspond with the size of a byte of data.

USB

USB stands for *Universal Serial Bus*. This is now the go-to standard for attaching things to devices.

At a basic level, USB is just a serial communications protocol at an agreed speed (which is negotiated when you first plug a USB device in). However the standard also defines *what data you send and when*, to ensure that things like computer mice, keyboards, and USB drives *just work* when you plug them in.

This applies to other types of devices, too. For instance, Espruino uses *USB CDC* or *Communications Device Class*. Many operating systems (Mac OS, Linux, Android, and Chrome OS) know about this and can communicate with Espruino without any drivers.

So let's have a go at sending some data from Espruino to our computer without using the USB connector.

Most computers have a connector for USB, but there are very few other connectors you can rely on, especially on smaller laptops. About the only other connector all computers have at the moment is a headphone jack, so maybe we can use that?

First, we're going to make a simple oscilloscope using our computer, so we can see what signals we're sending from Espruino.

Experiment 17: Making an Oscilloscope

You'll need the following to connect the oscilloscope (see Figure 12-4):

- An Espruino Pico board

- Breadboard

- A headphone extension lead with a 3.5mm jack plug (try and get one that has a single cable, one with a circular cross-section rather than a figure of eight)

- 2x 10k Ohm resistors

If your computer doesn't have a *line in* jack plug, you'll need to get a special cable that splits your headphone jack out into a headphone and microphone.

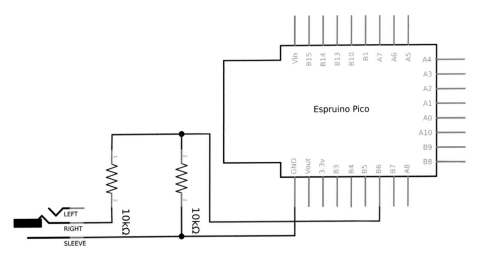

Figure 12-4 *Connecting the oscilloscope up*

1. First, put the Espruino board into the breadboard as before, with the USB connector over to the far left.

2. Strip back the end of the headphone cable (cutting the headphones off if there were some on it). It should look like this:

There should be two insulated wires, and some uninsulated shielding (which is the ground connection), as shown here:

If you have all uninsulated wires but they are different colors, each part of the braid has a very thin coating over it. You'll need to use a soldering iron and solder to melt through this and tin it so you can make a connection.

3. Twist the shielding together to stop it from getting tangled:

4. Strip the insulation off one of the insulated wires (leaving a centimeter of it), and twist the copper inside it together:

5. Twist the copper from the insulated wire onto one side of one resistor (we'll call it R1), and twist the copper from the shielding onto the other resistor (resistor R4), as shown here:

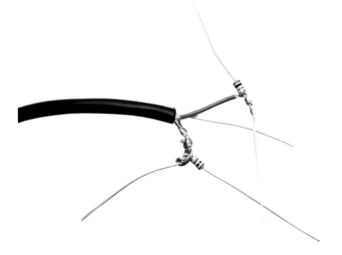

6. Put resistor R1 in the breadboard, with the end that has the twisted wire in a column just past the end of the Espruino Pico, and with the other end of the resistor on pin B6 (the third pin in from the end at the bottom).

7. Put resistor R2 in the breadboard, with the twisted wire end in GND (the left end of the Pico), and the other end in pin B6 with the other resistor. The finished breadboard should look like this:

You should now have your circuit! Give it a quick check and make sure it's connected properly. The only two wires to the Espruino Pico should go to GND and B6.

Using the Oscilloscope

1. Open the Chrome web browser.

2. Go to *https://espruino.github.io/webaudio-oscilloscope/* and make sure you allow it to use your microphone when prompted.

   The source code for the oscilloscope web page is available at GitHub (*https://github.com/espruino/webaudio-oscilloscope*).

3. Plug the headphone jack into your computer.

4. Open up the Sound Record settings on your computer and make sure the microphone is selected and the volume is set to *unamplified*. On many computers this will just be 100%.

5. Now connect the Espruino to USB, and enter the following code:

```
var t = 0;
setInterval(function() { Serial1.write(t); },5)
```

This will set it transmitting the value 0 at the default bit rate (baud rate) of 9600 baud, 20 times a second.

6. You should now see something like this on the screen:

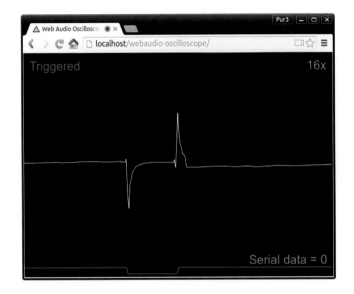

So what's happening here? Well, we're seeing the signal that the Espruino board's Serial Peripheral is creating to send one byte of data, in this case 0. It's complicated slightly because our *hacky* oscillocope's Line In input contains a capacitor that always tries to restore the input to zero volts (the middle of the oscilloscope trace). As a result, you only see blips when the signal changes state, not the current value.

To try to make it clearer, the oscilloscope page shows a blue line at the bottom, which is what it believes the initial input digital signal is.

You can set t to different values to send different things. For instance, here we've set t to 255:

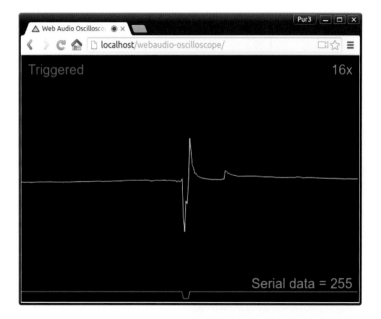

And here we've set `t` to `0x55`:

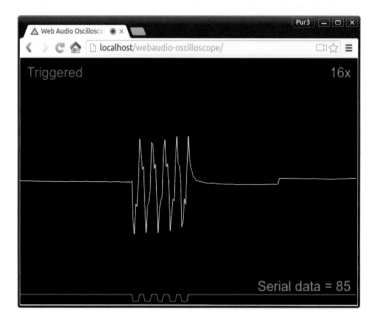

The following diagram shows what's actually being sent. When it's not doing anything, the wire is in the `1` state. There's a *start* bit (which is a 0) to show that something is about to be sent, then there are 8 bits of data, and finally there's a *stop* bit (which is a 1, and so can't actually be seen when we're sendding single characters).

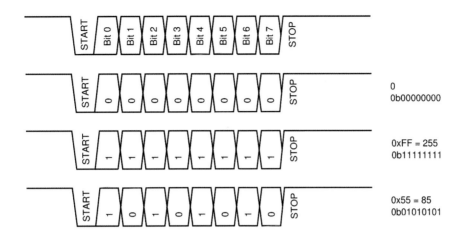

0
0b00000000

0xFF = 255
0b11111111

0x55 = 85
0b01010101

You should be able to see how the diagram matches up with the data that's being sent, and how the oscilloscope program is able to work out what the signal is and display it in the bottom right of the screen.

How does it decode it? In this case, all it does is wait for the signal to go from a 1 to a 0, indicating a *start* bit. It then looks at the signal again after a delay of 1.5 bits (1.5 / 9600 = 0.156ms) to get the first bit, and then again 1 bit (1.5 / 9600 = 0.104ms) later to get the second, and so on, as shown here:

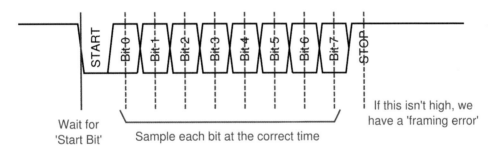

Wait for 'Start Bit'

Sample each bit at the correct time

If this isn't high, we have a 'framing error'

This code could be extended in order to read multiple characters and do something with them. In fact, you can use the same ideas to transmit information from your computer directly to the Espruino. There's some information on how to do this (and even how to program the Espruino with the Web IDE) on the Espruino website (*http://www.espruino.com/Headphone*).

Cutting the Cord: Infrared | 13

We've looked at how we can use the computer's microphone jack to look at the signals from the Espruino's serial port, but what if we want to communicate without wires? First, let's take a look at how we can use a PC's audio input to receive signals from an infrared remote control.

Experiment 18: Making the IR Receiver

You'll need:

- An IR receiver (HX1838, VS1838, TSOP348, or TSOP344) and remote control

- A battery or battery back supplying between 3 and 5 volts (a CR2032 watch battery in a holder is perfect)

- Breadboard

- The headphone extension lead with a 3.5mm jack plug from *"Experiment 17: Making an Oscilloscope"*

- 2x 10k Ohm resistors

Finding an IR Receiver

IR receivers and remote control combinations can easily be bought in kits from places like eBay and Amazon. Just search for keywords such as *Arduino IR Remote*; however, you can use pretty much any infrared remote control you might have lying around. You can also buy just the receiver seperately (the part number should be one of HX1838, VS1838, TSOP348, or TSOP344) or you can even scavenge it from an old bit of remote control equipment.

You can also buy TSOP322 and TSOP324 parts, but for these the pins are in a different order.

1. Connect everything up as shown here:

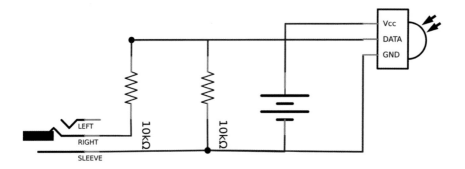

You can use the same headphone wiring that you used before, now you just need to connect it to the IR receiver and battery instead of the Espruino board. The connections for the IR receiver are as follows:

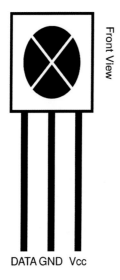

2. Your finished circuit should look a bit like this:

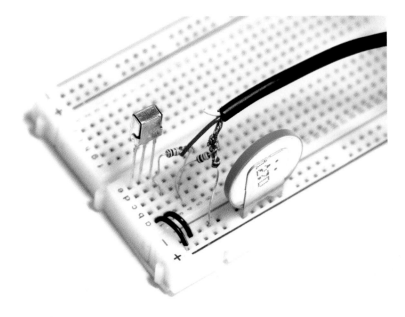

 Once wired up your battery will slowly run down, so be prepared to unplug the battery when you're finished.

Now you can use the oscilloscope from Chapter 12 (*http://bit.ly/2q8itQK*).

Normally, you should just see random noise on the wire. However, if you aim the remote control at the receiver and press a button, you should start to see some pulses appearing. IR remote controls don't transmit as quickly as the Espruino's serial did so they'll be more spaced out, but you should occasionally see things like this:

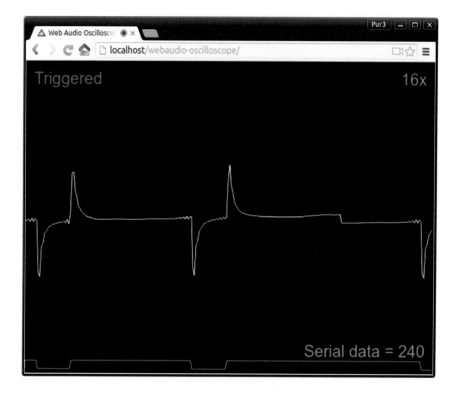

So what can we do with this? Last time we didn't write any code to run on our computer, but this time we'll write some JavaScript on a web page that will attempt to decode these signals.

Experiment 19: Decoding IR Signals

Here are our first steps to write the decoder:

1. Visit *https://jsfiddle.net*. You can use this to test out JavaScript code quickly and easily.

2. In the HTML area, type the following:

```
<body>
  <pre id="log"></pre>
</body>
```

3. In the JavaScript area, type:

```
var logElement = document.getElementById("log");
console.log = function(s) {
  logElement.innerHTML += s+"\n";
}
```

```javascript
// The threshold for detecting a change to a 1 or a 0 in the signal
var THRESH = 0.01;

// Do we think the input is a 1 or a 0?
var currentState = 0;
// How many samples have we been in this state (in samples)?
var timeInState = 0;

// Called when the input changes state
function changedState(newState, timePassed) {
  console.log((newState?"Lo":"Hi") + " for " + (time
Passed*1000).toFixed(2));
}

function processAudio(e) {
  var data = e.inputBuffer.getChannelData(0);
  // Now search for changes in value
  for (var i=0;i<data.length;i++) {
    // Did it suddenly go high? it's a 1
    if (currentState==0 && data[i]>THRESH) {
      currentState=1;
      changedState(1, timeInState / e.inputBuffer.sampleRate);
      timeInState = 0;
    }
    // Did it suddenly go low? it's a 0
    if (currentState==1 && data[i]<-THRESH) {
      currentState=0;
      changedState(0, timeInState / e.inputBuffer.sampleRate)
      timeInState = 0;
    }
    timeInState++;
  }
}

function startRecord() {
  window.AudioContext = window.AudioContext || window.webkitAudioContext;
  if (!window.AudioContext) {
    console.log("No window.AudioContext");
    return; // no audio available
  }
  navigator.getUserMedia = navigator.getUserMedia ||
                           navigator.webkitGetUserMedia ||
                           navigator.mozGetUserMedia;
  if (!navigator.getUserMedia) {
    console.log("No navigator.getUserMedia");
    return; // no audio available
  }

  var context = new AudioContext();
  var userMediaStream;
  var inputNode = context.createScriptProcessor(4096, 1/*in*/, 1/*out*/);
```

```
            window.dontGarbageCollectMePlease = inputNode;
            inputNode.onaudioprocess = processAudio;

            navigator.getUserMedia({
                video:false,
                audio:{
                  mandatory:[],
                  optional:[{ echoCancellation:false },
                    { googEchoCancellation: false },
                    { googAutoGainControl: false },
                    { googNoiseSuppression: false },
                    { googHighpassFilter: false }
                   ,{ sampleRate:22050 /* 44100 */ }]
                }
            }, function(stream) {
                var inputStream = context.createMediaStreamSource(stream);
                inputStream.connect(inputNode);
                inputNode.connect(context.destination);
                console.log("Record start successful");
            }, function(e) {
                console.log('getUserMedia error', e);
            });
        }

        startRecord();
```

There's a lot to this, but the majority of the code makes sure that the processAudio function keeps getting called with blocks of sound data that have been recorded. processAudio itself then attempts to work out whether the signal is currently a digital 1 or a 0 (this is like the blue line on the oscilloscope web page), and it then calls the changedState function whenever it thinks the state has changed. The changedState function then just prints out how long it thinks each pulse was (in milliseconds).

4. Click Run in the top left, and the web browser should display Record start successful in the bottom-right pane.

5. Take your remote control and press a key. You should now see something like this:

```
Record start successful
Hi for 2561.09
Lo for 8.87
Hi for 4.49
Lo for 4.94
Hi for 0.63
Lo for 0.48
Hi for 0.61
Lo for 0.50
Hi for 0.61
Lo for 0.50
Hi for 0.61
```

```
Lo for 0.50
Hi for 1.72
Lo for 0.50
Hi for 1.75
Lo for 0.50
Hi for 1.75
Lo for 0.48
Hi for 1.75
...
```

 If you don't get a list like this one, try changing the THRESH *value. Lower values will be more sensitive, and higher will be less.*

It'll depend on the type of remote control you have, but you should start to see a pattern. In this case, there's the first very long pause (between clicking Run and pressing a button on the remote control), followed by a long pulse of IR light. These IR receivers output a 1 when no IR signal is present, and output 0 when the remote control's signal is detected.

After the initial long pulse, all of the lines marked Lo show times of around 0.5ms, whereas the lines marked Hi are either around 0.6ms or 1.7ms. This means that the time between pulses is used to store the data: long pulses will be a 1, and short will be a 0 (or vice versa). Your remote control will almost certainly be different (as remote controls vary by manufacturer), but should be similar. Either the time the pulse is on (Lo), or the time it is off (Hi), will vary between being long or short.

6. So now, we can try to decode it. We'll just change the changedState function:

```
var currentCode = "";

function gotCode(code) {
  console.log("Got code "+code);
}

// Called when the input changes state
function changedState(newState, timePassed) {
  if (timePassed > 0.02) {
    // Let's assume there was a gap. Handle the last code, and reset it
    if (currentCode!="") gotCode(currentCode);
    currentCode = "";
  }
  // Add a bit only when the signal was high (it's now 0)
  // since in the last code, 'Hi' was the signal
  // that changed while 'Lo' was constant
  if (newState == 0) {
    if (timePassed > 0.001) currentCode += "1";
    else currentCode += "0";
```

```
        }
    }
```

7. If you use the remote control, you should now get something like:

```
Got code 110000000011111111101000101001011101
Got code 111110000001110000000010000011011111
Got code 111110000001110000000010000011011111
Got code 111110000001110000010100000001011111
Got code 111110000001110000000110000010011111
Got code 111110000001110000000110000010011111
Got code 111110000001110000000001000011101111
Got code 111110000001110000000001000011101111
Got code 111110000001110000010010000001101111
Got code 111110000001110000010010000001101111
Got code 111110000001110000010010000001101111
Got code 111110000001110000010010000001101111
Got code 111110000001110000010100000101011111
Got code 111110000001110000010100000101011111
Got code 111110000001110000010100000101011111
Got code 111110000001110000010010000001101111
```

You'll get a different code for each button that you press. If you keep a button held, you might get the same repeated code; however, on some other remote controls you might just get a short code like 11.

 Not getting codes of the correct length? Experiment by changing the value of THRESH. You may well get an occasional corrupted code that isn't the correct length (especially if the remote control is some way away or pointing in the wrong direction), but this is to be expected.

You might notice that the first time you press a button on the remote, you don't get anything printed. In fact, when you press a button, you'll be getting the code for the *last* button reported. This is because gotCode is only called when the signal next changes after a delay. If you just press the button once, the code is recorded, but since the signal doesn't change again, it isn't reported until the next IR signal is received.

8. To get around this, we need to make sure we call gotCode if the signal hasn't changed after a certain amount of time:

Add the following code right at the end of the processAudio function:

```
if (timeInState > 2000) {
    if (currentCode!="") gotCode(currentCode);
    currentCode = "";
}
```

Now when you use the remote control, you should get the signal reported correctly as soon as the button is pressed.

Experiment 20: Using Our Decoded Signal

So now our function `gotCode` is called whenever we receive a code from the remote control. We can easily copy the values that we printed and do something depending on the value, like changing something onscreen.

1. Modify the HTML to the following:

    ```
    <body>
      <div id="box" style="width:100px;height:100px"></div>
      <pre id="log"></pre>
    </body>
    ```

    This will add a fixed-size box at the top of the page.

2. Next, we're going to modify `gotCode` so that it changes the color of the box when a button is pressed.

    Decide on four buttons to use, press them, and copy the code that is reported into the following code. For instance, I'm using the buttons marked 1 to 4:

    ```
    function gotCode(code) {
      if (code=="111110000011100000010000011011111")
        document.getElementById("box").style.background="red";
      if (code=="111110000011100001010000001011111")
        document.getElementById("box").style.background="green";
      if (code=="111110000011100000110000010011111")
        document.getElementById("box").style.background="yellow";
      if (code=="111110000011100000001000011101111")
        document.getElementById("box").style.background="blue";
      console.log("Got code "+code);
    }
    ```

    Now, when you press the buttons on the remote control, you can change the color of the square on the screen.

You could easily use this functionality to control a game or quiz that was written as a web page.

Experiment 21: Using Our Remote Control on the Net, with dweet.io

However, web pages can request new web pages all by themselves, so you could use this functionality to navigate between pages. You can even request a page that is able to give information to a web server.

For this example, we'll use dweet.io—an amazingly useful website that allows you to simply push data, and to then read it back from another place. On the front page of dweet.io there's more information about exactly how to use it, but all we need to do is to request a certain web page, and that will be enough to store information on dweet.io's servers.

1. Think of a random name. For this example I'll just use `espruino`.

2. Navigate to *http://dweet.io/follow/espruino* in your browser, changing *espruino* to the name you came up with.

   Most likely, dweet.io will report:

   ```
   Sorry this thing isn't a thing
   ```

3. Now, in another window, navigate to *https://dweet.io/dweet/for/espruino?code=hello* (again, changing *espruino*). This will write the value `hello` to the attribute `code` on dweet.io.

4. Now refresh the *http://dweet.io/follow/espruino* window. It should report `code` being `hello`:

5. In the *https://dweet.io/dweet/for/espruino?code=hello* window, change the URL to *https://dweet.io/dweet/for/espruino?code=testing* and you should see that the

other window now automatically updates. You can even open the *http://dweet.io/follow/* window on a new device and it will still work.

We can now control this automatically from our IR receiver!

6. Change the `gotCode` function to the following, changing the `espruino` text in the URL to the name that you came up with:

```
// the last code we received
var lastCode;

function gotCode(code) {
  console.log("Got code "+code);

  // Make sure we only call this when we get a new code
  if (lastCode == code || code.length<20) return;
  lastCode = code;

  // Now send a message to dweet.io
  var oReq = new XMLHttpRequest();
  oReq.addEventListener("load", function() {
    console.log("Got response: "+this.responseText)
  });
  oReq.open("POST", "https://dweet.io/dweet/for/espruino?code=_"+code);
  oReq.send();
}
```

If you use the remote control now, you should see your *http://dweet.io/follow/* window updating with the code from the remote control!

 Because dweet.io is a free service, it is rate-limited. You can only send one update every few seconds. If you press a few keys very quickly you may find that new keys are getting ignored.

Experiment 22: Using Our Remote Control on the Net, with IFTTT

dweet.io is a good way to test code, and it's pretty handy if you're able to write more software that can read the latest values for your device. However, it doesn't really integrate with any other web services.

If you want something that does, you can use If This Then That (*https://ifttt.com*), a website that lets you create a set of the rules of the form *If this happens, do that*. While IFTTT usually only deals with other commercial products, they have added a new *channel*, called the Maker Channel (*https://ifttt.com/maker*).

This allows you to use it in much the same way as you did with dweet.io, so let's set it up such that when you press a specific button on your remote control it will send you an e-mail.

1. Navigate to *https://ifttt.com* and sign up for an account.

2. Now go to *https://ifttt.com/maker_webhooks* and click *Connect*. On the next page, click *Documentation* and you should see a key value:

Your key is:

◄ Back to service

To trigger an Event

Make a POST or GET web request to:

```
https://maker.ifttt.com/trigger/ {event} /with/key/
```

With an optional JSON body of:

```
{ "value1" : "      ", "value2" : "      ", "value3" : "      " }
```

The data is completely optional, and you can also pass value1, value2, and value3 as query parameters or form variables. This content will be passed on to the Action in your Recipe.

You can also try it with curl from a command line.

```
curl -X POST https://maker.ifttt.com/trigger/{event}/with/key/
```

Test It

3. Replace the gotCode function with the following, and replace xxx_MY_KEY_HERE_xxx with the long series of digits shown as your key on the maker page:

```
// the last code we received
var lastCode;

function gotCode(code) {
  console.log("Got code "+code);

  // Make sure we only call this when we get a new code
```

```
  if (lastCode == code || code.length<20) return;
  lastCode = code;

  // Now send a message to IFTTT
  var oReq = new XMLHttpRequest();
  oReq.addEventListener("load", function() {
    console.log("Got response: "+this.responseText)
  });
  oReq.open("POST", "https://maker.ifttt.com/trigger/infrared/with/key/"+
                    "xxx_MY_KEY_HERE_xxx"+
                    "?value1=_"+code);
  oReq.send();
}
```

4. Now click on your login name in the top right and click *New Applet*.

5. Click *this* and then choose *Maker Webhooks*.

6. Select "Receive a web request" and type **infrared** into the *Event Name* box, then click *Create Trigger*.

 infrared *is the same name that we used in the URL in* gotCode *earlier.*

7. Now click *that*, choose *Email*, *Send me an email*, and then click *Create Action*.

You should now have something that looks like the following:

8. Click *Create Recipe*.

9. Now you're ready to go. Just press a button on the remote control, and you should be sent an email!

HTTP Access Control

JavaScript on web browsers has something called HTTP access control (*http://bit.ly/mdn-cors*). This restricts the type of HTTP accesses that can be made from a web page to a totally different web server for security, unless the server opts out. While dweet.io opts out, unfortunately IFTTT doesn't (yet). This means that while the preceding code works fine (as the initial HTTP request has been made by the time the web browser figures out there is a problem), you will get an error reported in your web browser's console each time a HTTP request is made.

IFTTT can do much more than send emails. You can send tweets, make notes, add rows to a spreadsheet on Google Drive, or even control some web-enabled devices such as mains sockets. Play around and see what you can do!

While we've used JSFiddle here to make it easy for you to change your code quickly, you can easily host this code on your own website, or you can use services such as GitHub pages (*https://pages.github.com*) to permanently host your work. Just bear in mind that in the case of IFTTT your API key may be visible to everyone, meaning that anyone who wants to could trigger your IFTTT rules!

The `getUserMedia` *functionality used here is also supported on Android. If you have an old mobile phone or tablet, you can set it up in the same way and use it to relay commands to the internet!*

Cutting the Cord: Radio Signals

While infrared is nice and simple, there are some real drawbacks. For instance, the low data rate and the need to have *line of sight* to the receiver.

The next step is to use proper radio transmission. There are plenty of different types of radio transmission for data: there's Bluetooth, Bluetooth Low Energy (sometimes known as Bluetooth Smart or BLE), Zigbee, WiFi, LoRa, and many others.

While very powerful, most of these systems use complicated radio transceiver chips. However, there are still some radios that are extremely simple—for example, the kind often used by cheap wireless sensors or wireless doorbells.

These use simple AM transmitters and receivers (see Figure 14-1) that transmit in the free radio bands: 315MHz (USA) and 433MHz (European). Much like the infrared remote controls, the protocol used varies depending on the manufacturer. We'll use those for this experiment, as it gives you the ability to see the signals that are being sent and received.

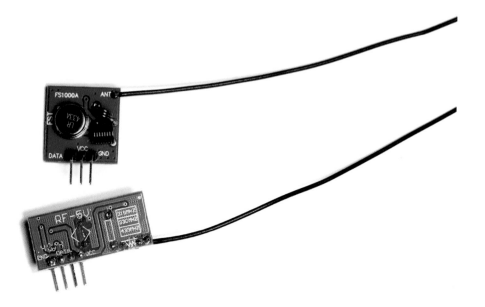

Figure 14-1 *Common radio transmitter (top) and receiver (bottom) modules*

Experiment 23: Wiring Up the Receiver

You'll need:

- A 315Mhz (USA) or 433Mhz (Europe) radio receiver

- A length of insulated, solid core wire

- A battery or battery pack supplying between 3 and 5 volts (a CR2032 watch battery in a holder is perfect)

- Breadboard

- The headphone extension lead with a 3.5mm jack plug from "Experiment 18: Making the IR Receiver"

- 2x 10k Ohm resistors

You'll need to create the circuit shown in Figure 14-2. This is almost identical to the circuit you made for the IR receiver in "Experiment 18: Making the IR Receiver".

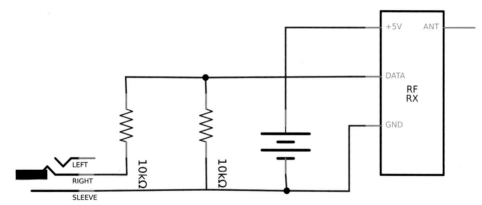

Figure 14-2 *Connecting the receiver*

1. Cut the solid core wire into two lengths for the antenna, depending on which reciever you have:

315MHz	23.8 cm
433MHz	17.3 cm

2. If you have a soldering iron, solder one antenna onto the receiver where it is marked, otherwise just strip the wire, poke it through, and twist it to hold it on.

3. Connect the wire and two resistors as you did for the headphone oscilloscope in "Experiment 17: Making an Oscilloscope".

    The finished receiver should look something like this:

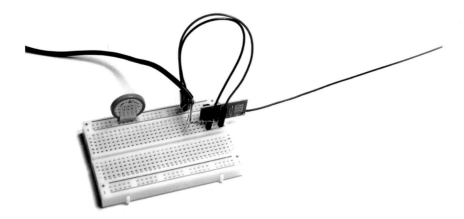

You can now test it by running the WebAudio oscilloscope (*https://espruino.github.io/webaudio-oscilloscope/*) that we used before.

As mentioned, the 315Mhz and 433Mhz bands are in use by all kinds of devices. They're also relatively long range, with even low-power transmitters managing 100m or so. As a result, you'll find that the band is relatively congested. You might find just looking at the output of the receiver that it's receiving quite a lot of data just sitting there. This is what I see when viewing the web page:

If you do have a device that works on the same band (like a wireless remote or doorbell) try activating it and see if you can detect anything.

Experiment 24: Wiring Up a Transmitter

Now that we've got a receiver, let's wire up a transmitter. We'll be re-creating the circuit in Figure 14-3.

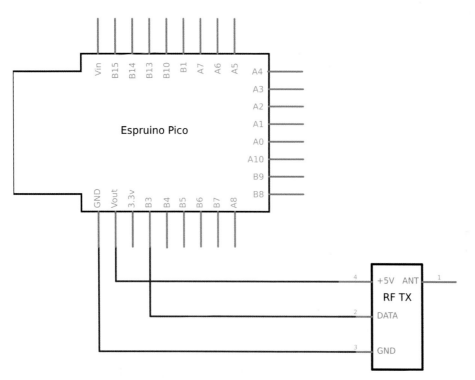

Figure 14-3 *How to wire an Espruino Pico to a radio transmitter*

This is nice and easy! You'll need:

- A 315Mhz (USA) or 433Mhz (Europe) radio transmitter
- A length of insulated, solid core wire
- Breadboard
- Three patch wires
- An Espruino Pico

To wire the transmitter, simply:

1. Add the antenna as you did for the receiver.

2. Put the Pico and the transmitter in the breadboard, and wire them up so it looks as follows:

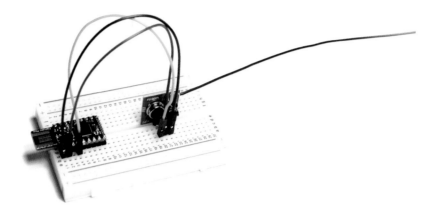

Experiment 25: Transmitting from Espruino

Now we're ready to start sending data directly from Espruino to your computer.

First, we're just going to create a test signal: a long series of pulses. On for 1 millisecond, and off for 2.

To do this, we can send an array of pulse lengths to the `digitalPulse` command as we did for the Baird TV. As before, we need an odd number of elements in the array so that after all pulses have been sent the transmitter is off. To solve this we'll just create the array with one element, and then add two elements at a time later on.

1. Enter the following code on the lefthand side of the Web IDE:

   ```
   var d = [1 /* on */]
   for (var i=0;i<100;i++) d.push(2 /* off */, 1 /* on */);
   ```

2. Now we can actually send our test signal:

   ```
   digitalPulse(B3,1,d)
   ```

   You should see, for a fraction of a second, something like this:

3. So now we can look at sending some data, but we need a way to distinguish it from all the other signals and radio noise that we're receiving.

   We're going to send a long, 5-millisecond pulse at the start. This has the benefit that it teaches our radio receiver how high the signal strength for a 1 should be. After this, the radio receiver will start ignoring any radio signals that are significantly less than this.

   After this, we'll send pulses of differing lengths. Just like with infrared, a long pulse will be a 1, and a shorter pulse will be a 0. We'll grab the data we want to send from a string of text.

   Each character is 8 bits, so we'll send 8 pulses for each character.

   Finally, we'll send a 3-millisecond end pulse to indicate the end of the received data (otherwise the receiver might accidentally end up reading random radio signals after the end of our transmission).

   Enter the following code on the righthand side of the IDE and click *Upload*:

   ```
   function transmit(txt) {
     // Ensure we're dealing with a string
     txt = txt.toString();
     // Work out what to send
     // Initial 5ms pulse
   ```

```
var d = [5];
// data for each character
for (var i=0;i<txt.length;i++) {
  var ch = txt.charCodeAt(i);
  for (var j=0;j<8;j++) {
    d.push(1, (ch&128)?1.5:0.5);
    ch<<=1;
  }
}
// Finally add a 3ms 'finish' pulse
d.push(1,3);
// Send it
digitalPulse(B3,1,d);
}
```

4. Now type the following on the lefthand side of the IDE:

```
transmit("Hello")
```

This will send the text Hello. You should see some data flash up on the oscilloscope, but how do we decode it?

Experiment 26: Decoding the Received Data

We're going to start off with the same audio recording code we had for the infrared receiver.

1. Open Chrome and go to *http://jsfiddle.net*.

2. Add the following code under the HTML heading:

```
<body>
  <pre id="log"></pre>
</body>
```

3. Add the following under the JavaScript heading:

```
var logElement = document.getElementById("log");
console.log = function(s) {
  logElement.innerHTML += s+"\n";
}

// The threshold for detecting a change to a 1 or a 0 in the signal
var THRESH = 0.01;

// Do we think the input is a 1 or a 0?
var currentState = 0;
// How many samples have we been in this state (in samples)?
var timeInState = 0;

// The data we've received so far
var currentCode;
```

```
// Called when the input changes state
function changedState(newState, timePassed) {
  console.log((newState?"Lo":"Hi") + " for " + (time
Passed*1000).toFixed(2));
}

function processAudio(e) {
  var data = e.inputBuffer.getChannelData(0);
  // Now search for changes in value
  for (var i=0;i<data.length;i++) {
    // Did it suddenly go high? it's a 1
    if (currentState==0 && data[i]>THRESH) {
      currentState=1;
      changedState(1, timeInState / e.inputBuffer.sampleRate);
      timeInState = 0;
    }
    // Did it suddenly go low? it's a 0
    if (currentState==1 && data[i]<-THRESH) {
      currentState=0;
      changedState(0, timeInState / e.inputBuffer.sampleRate)
      timeInState = 0;
    }
    timeInState++;
  }
}

function startRecord() {
  window.AudioContext = window.AudioContext || window.webkitAudioContext;
  if (!window.AudioContext) {
    console.log("No window.AudioContext");
    return; // no audio available
  }
  navigator.getUserMedia = navigator.getUserMedia ||
                           navigator.webkitGetUserMedia ||
                           navigator.mozGetUserMedia;
  if (!navigator.getUserMedia) {
    console.log("No navigator.getUserMedia");
    return; // no audio available
  }

  var context = new AudioContext();
  var userMediaStream;
  var inputNode = context.createScriptProcessor(4096, 1/*in*/, 1/*out*/);
  window.dontGarbageCollectMePlease = inputNode;
  inputNode.onaudioprocess = processAudio;

  navigator.getUserMedia({
      video:false,
      audio:{
        mandatory:[],
```

```
            optional:[{ echoCancellation:false },
               { googEchoCancellation: false },
               { googAutoGainControl: false },
               { googNoiseSuppression: false },
               { googHighpassFilter: false }
              ,{ sampleRate:22050 /* 44100 */ }]
            }
       }, function(stream) {
          var inputStream = context.createMediaStreamSource(stream);
          inputStream.connect(inputNode);
          inputNode.connect(context.destination);
          console.log("Record start successful");
       }, function(e) {
          console.log('getUserMedia error', e);
       });
    }

    startRecord();
```

Don't click Run just yet, because unless you are in a very rural area you're going to have a huge amount of data printed to the screen, as every single bit of data received will be output!

4. Instead, replace `changedState` with some code that decodes the sent signal. It will:

   - Wait until a (roughly) 5ms pulse is found.

   - Start recording the bits of data is receives (unless they're obviously the wrong length).

   - Call `gotData` with the data received when it gets a 3ms end pulse.

   - Use `gotData` to reassemble the data into a `String`.

```
// The data we've received so far, or 'undefined' if we are not
// receiving any data at the moment
var currentCode;

function gotData(data) {
  // Print the raw data
  console.log(data);
  // Reconstruct a String from this
  var str = "";
  for (var i=0;i<data.length;i+=8) {
     str += String.fromCharCode(parseInt(data.substr(i,8),2));
  }
  console.log("   " + JSON.stringify(str));
}

// Called when the input changes state
function changedState(newState, timePassed) {
```

```
  if (newState!=0) return;
  var ms = timePassed * 1000; // time in milliseconds
  // Check the pulse length
  if (ms > 4.5 && ms < 5.5) {
    // It's a 5ms 'start' pulse
    currentCode = "";
  } else if (ms > 2.5 && ms < 3.5) {
    // It's a 3ms 'end' pulse
    if (currentCode!==undefined && currentCode.length)
      gotData(currentCode);
    currentCode = undefined;
  } else if (ms<2) {
    // It's hopefully a data pulse
    if (currentCode!==undefined)
      currentCode += (ms>1) ? "1" : "0";
  } else {
    // unknown pulse size - give up on receiving
    currentCode = undefined;
  }
}
```

5. Now click Run.

6. In the Web IDE, press the ⬆ to reselect `transmit("Hello")`, and hit `Enter` to send it again.

   You may have to do this more than once, but hopefully eventually you'll receive the data from the Espruino board inside JSFiddle.

7. As with all radios, there's no guarantee that your signal won't be interrupted by something else, or that some other device's signal won't get interpreted as a legitimate signal by your receiver. It's easy to end up with the wrong data or nothing at all, so most systems perform some sanity checks on the data received.

   For example, a simple test would be to ensure that the data received was a multiple of 8 bits long:

   Add the following to the very start of `gotData`:

   ```
   // Return if data is not a multiple of 8 bits long
   if (data.length & 7) return;
   ```

   In reality more complex checks are performed, such as *Cyclic Redundancy Checks* (CRCs). CRCs produce a number based on the contents of the message, and that number is sent as well as the message. If it doesn't match, the message is considered to be corrupt. For example, with an 8-bit CRC there are 256 possible combinations, so if the CRC matches there is now a less than 0.5% chance that it is wrong.

   Sending `"Hello"` wasn't desperately useful. What if we wanted to send the current temperature of the Espruino board?

8. Enter the following code in the Web IDE, which will send the temperature every 5 seconds. `toFixed` takes the floating-point number, and converts it to a string with the given number of decimal places. For example: `(27.87103308575).toFixed(2) == "27.87"`.

```
setInterval(function() {
   transmit(E.getTemperature().toFixed(2));
}, 5000);
```

After a few seconds you should start to see temperature readings arriving!

You could now put the data you received online using the examples from the previous chapter. For example, you could change `gotData` in JSFiddle to the following:

```
function gotData(data) {
  // Return if data is not a multiple of 8 bits long
  if (data.length & 7) return;
  // Print the raw data
  console.log(data);
  // Reconstruct a String from this
  var str = "";
  for (var i=0;i<data.length;i+=8) {
    str += String.fromCharCode(parseInt(data.substr(i,8),2));
  }
  console.log("   " + JSON.stringify(str));
  // Now send a message to dweet.io
  var oReq = new XMLHttpRequest();
  oReq.addEventListener("load", function() {
    console.log("Got response: "+this.responseText)
  });
  oReq.open("POST","https://dweet.io/dweet/for/espruino_tmp?temp="+str);
  oReq.send();
}
```

You can then go to *http://dweet.io/follow/espruino_tmp*, and if you leave the window open you can even see a graph of the current temperature building up over time (as before, you ought to use your own ID, because if two people use `espruino_tmp` at the same time, both of their data uploads will be recorded!):

Was the Message Received?

For this experiment, the Espruino board only transmits data. It can't ever know if the data it sent was received or not. Many devices in the 315/433 Mhz bands work like this. They transmit the same data multiple times, just to make it more likely that it will arrive.

If both ends can transmit and receive then the *receiver* can send a special acknowledgment message when it gets some data. If the sender doesn't get an acknowledgment, it can resend the data, making it much more likely that it arrives.

However, in many cases (such as a room temperature monitor), it isn't vital that data is received. If one message is missed, hopefully there will just be a new one in a few seconds!

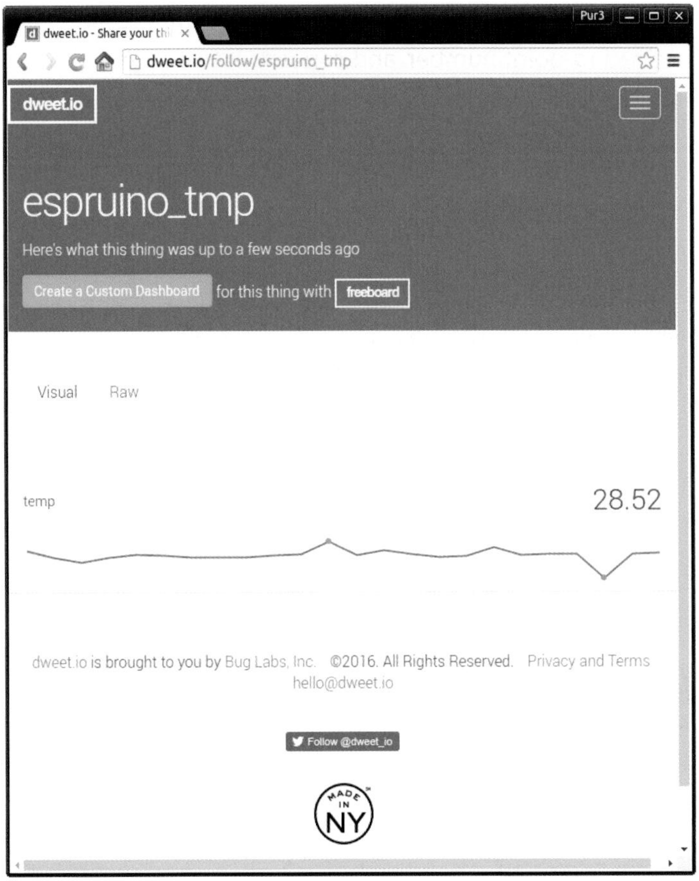
dweet.io - Share your th
dweet.io/follow/espruino_tmp
dweet.io
espruino_tmp
Here's what this thing was up to a few seconds ago
Create a Custom Dashboard
for this thing with
freeboard
Visual
Raw
temp
28.52
dweet.io is brought to you by Bug Labs, Inc. ©2016. All Rights Reserved. Privacy and Terms
hello@dweet.io
Follow @dweet_io

Connecting with WiFi | 15

The options we've looked at so far have involved having a *bridge* to connect a device to the internet. In the infrared and RF modules that bridge has been your PC or an Android phone.

In many cases that may not be what you want. Often you'll need a device that connects directly to your WiFi network, without a separate bridge.

To do this, you're going to need something to handle the WiFi connection for you. There are many different modules around, but at the moment the cheapest, most easily sourceable modules are based on the ESP8266 (Figure 15-1). The ESP8266 is made by EspressIF (*https://espressif.com*). It's a great little chip that contains quite a powerful microcontroller as well as all the 2.4GHz radio electronics needed for WiFi communications. The chip itself needs external flash memory (the little 8-pinned chip by the side) as well as a tuned aerial, so the modules are by far the easiest way to buy and use the ESP8266.

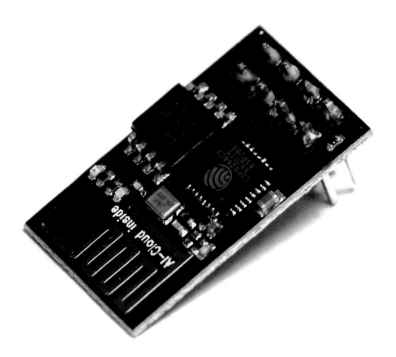

Figure 15-1 *The ESP8266 ESP01 module*

The AT Command Set

While the microcontroller in the ESP8266 can run any code, by default it comes pre-installed with an *AT command* firmware. AT stands for *ATtention*. Any commands that need to be sent to the ESP8266 start with the characters AT, which helps to differentiate them from data.

The *AT*, or *Hayes*, command set was created back in 1981 for the Hayes Smartmodem. Variants of it have since been used in all kinds of devices, most notably in the modems of most mobile telephones. Manufacturers tend to add their own *AT* commands so there is no guarantee that one device will work with another's drivers. The *AT* command set merely tells you that the commands will start with *AT*, not what those commands will be!

While we're showing you how to connect an ESP8266 board to a Pico here (as that's used for the other chapters), you can also use an Espruino WiFi board (*http://www.espruino.com/WiFi*), which has an ESP8266 module pre-installed.

Espruino WiFi

If you don't want to have to connect up separate hardware for WiFi, you can buy an Espruino WiFi board. This is effectively a more powerful Espruino Pico with an ESP8266 module pre-installed.

There's more information on the board at the Espruino website (*http://www.espruino.com/WiFi*). There will be other notes in this chapter where instructions differ if you're using an Espruino WiFi board.

Experiment 27: Adding WiFi to Your Pico

First, you'll need to physically connect the ESP8266 module to the Espruino Pico.

You'll need:

- An ESP01 ESP8266 module
- Breadboard
- Five male-to-female patch wires
- An Espruino Pico

Wiring is easy, just connect the five wires as shown in Table 15-1.

Table 15-1 *ESP8266 to Espruino Pico connections*

ESP8266	Espruino Pico
GND	GND
RX	B6
TX	B7
CH_PD	3.3v
VCC	3.3v

Figures 15-2 and 15-3 show where the pins are on the ESP01 module.

Figure 15-3 *A top view of the ESP01 module*

Figure 15-3 contains the actual pin connections.

Top View

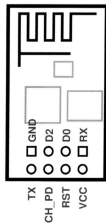

Figure 15-4 *The connections of the ESP01 module*

Your finished wiring should look like Figures 15-4 and 15-5.

Figure 15-5 *The five wires connected to an ESP01 module*

Figure 15-6 *The five wires connected to the Espruino Pico*

Try not to use wires that are too long. The ESP8266 can draw a lot of power (over 200mA) when connecting to a WiFi network, and long wires on the power lines can cause it to be unreliable.

Making This Tidier

The ESP8266 ESP01 module is a bit tricky to mount on breadboard because of the block of 4×2 pins on it.

However, if you have a soldering iron handy then you can just cut the middle four pins off, and then short the CH_PD pin to 3.3v on the module (Figure 15-6).

Figure 15-7 *A modified ESP01 module that only needs four wires connected*

You can then place it straight into the breadboard and connect with only four wires, as shown in Figure 15-7.

Figure 15-8 *Using a modified ESP01 module with an Espruino Pico board*

Various suppliers also provide more breadboard-friendly breakout boards for ESP8266 modules.

You can even buy small *shims* that allow you to attach the ESP8266 module (*http://www.espruino.com/ESP8266*) directly to your Pico, in a nice compact package (Figure 15-8).

Figure 15-9 *Using a "shim" to add WiFi to your Espruino Pico in a very compact way*

All in One!

Even though we're connecting an Espruino board to an ESP8266 module (*http://www.espruino.com/EspruinoESP8266*), the ESP8266 module used contains a fully programmable microcontroller all of its own. It is possible to run Espruino directly on the ESP8266 module without the need for an Espruino board.

This does work well, but the microcontroller in the ESP8266 has to give WiFi communications priority (and doesn't have the wealth of peripherals that Espruino boards have), so life won't be quite as easy as if you used the ESP8266 and a separate microcontroller.

Experiment 28: Testing Your Wiring

The next step is to check that your module is connected and working (there's no need to do this on the Espruino WiFi board).

1. Enter the following code on the righthand side of the Web IDE and click *Upload*:

```
var l="";
Serial1.on('data', function(d) {l+=d;});
Serial1.setup(115200, { tx: B6, rx : B7 });
Serial1.write("AT+GMR\r\n");
setTimeout(function(){console.log(l);},1000);
```

This code simply sends the `AT+GMR` command and returns the response. `AT+GMR` is a sort of *Who are you?* command that causes the ESP8266 to print the version of its software.

After a second, you will hopefully get something a bit like this:

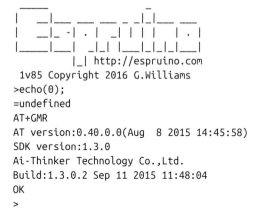

```
 1v85 Copyright 2016 G.Williams
>echo(0);
=undefined
AT+GMR
AT version:0.40.0.0(Aug  8 2015 14:45:58)
SDK version:1.3.0
Ai-Thinker Technology Co.,Ltd.
Build:1.3.0.2 Sep 11 2015 11:48:04
OK
>
```

If nothing is, or you have an AT version of less than `0.25.0.0`, head over to the Espruino website (*http://www.espruino.com/ESP8266*) for some troubleshooting code to help you out.

Experiment 29: Connecting to WiFi

Now we're ready to connect to WiFi!

1. Copy the following code on the righthand side of the Web IDE, and change `WIFI_NAME` and `WIFI_KEY` to the name and password of your WiFi network:

```
var WIFI_NAME = "";
var WIFI_KEY = "";

function onConnected() {
  console.log("Connected");
}

Serial1.setup(115200, { tx: B6, rx : B7 });
var wifi = require("ESP8266WiFi_0v25").connect(Serial1, function(err) {
  if (err) throw err;
  console.log("Connecting to WiFi");
  wifi.connect(WIFI_NAME, WIFI_KEY, function(err) {
    if (err) {
      console.log("Connection error: "+err);
      return;
    }
    onConnected();
```

```
  });
});
```

 Espruino WiFi

On the Espruino WiFi board, the code you use should look like this:

```
var WIFI_NAME = "";
var WIFI_KEY = "";

function onConnected() {
  console.log("Connected");
}

var wifi = require("EspruinoWiFi");
console.log("Connecting to WiFi");
wifi.connect(WIFI_NAME, { password : WIFI_KEY }, func
tion(err) {
  if (err) {
    console.log("Connection error: "+err);
    return;
  }
  onConnected();
});
```

It's almost the same, apart from the code that initializes the WiFi at the beginning.

If all goes well, you'll get something like this:

```
Connecting to WiFi
Connected
```

Did you get Uncaught WiFi connect failed: FAIL ? If so, recheck your WiFi details. You might also need to make sure you're well within range (the ESP8266's WiFi range isn't quite as good as your laptop's).

Now we're ready to connect to the internet.

2. On the lefthand side of the IDE, enter the following:

```
require("http").get("http://www.pur3.co.uk/hello.txt", function(res) {
  res.on('data', print);
});
```

We upload code using the right side of the IDE, but enter individual commands on the left side because we can do things without having to wait for WiFi to reconnect, which we would have to do if we re-uploaded code.

This will request a web page that simply says `Hello World!`. The function is called when a connection is made, and `res.on('data', print);` ensures that the `print` function is called whenever more data is received.

Often we'll want to deal with data as it comes in over WiFi rather than saving it all into one variable, as it wouldn't be hard for a web server to send us files that are so big that they would fill up all our memory!

Experiment 30: Sending Data to the Internet

Now, let's do something more interesting. In Chapter 14, we used dweet.io to send the temperature. How would we do the same thing here?

1. Enter the following code on the lefthand side of the IDE (after you've used the code in the last experiment):

```
function sendDweet() {
  var str = E.getTemperature().toFixed(2);
  var url = "http://dweet.io/dweet/for/espruino_tmp?temp="+str;
  require("http").get(url, function(res) { });
}

setInterval(sendDweet, 10000);
```

2. If you now go to *http://dweet.io/follow/espruino_tmp* you should see the temperature of your Espruino board updating once every 10 seconds! As before, you should probably change `espruino_tmp` to something else or you'll start seeing other people's data!

So that was nice and easy. We did cheat a little to keep this nice and simple, though. To update the dweet.io service in the correct way, we should really do an HTTP POST request (not a GET). There are some better examples of how to do this at the Espruino website (*http://www.espruino.com/IoT+Services*).

Why Shouldn't We Use a GET?

When you do an HTTP GET, the web server (and any proxy servers in the way) assume that you are only trying to retrieve data, not send it. As a result, if they have a *cached* page available that matches the description then they will just give you that cached page, and will never contact the server. Your data will be lost.

HTTP POST requests mean that you intend to send the server some information. They'll always be passed straight through, ensuring your data gets there intact.

Finally, we should look at making our code work straight after the Pico is powered on. So how do we do that?

3. Well, when the Pico first starts, it calls a function called `onInit` if it exists. In that function, we need to make sure we initialize the ESP8266 WiFi and start connecting. We'll just put all our WiFi connection code in there.

4. Next, we only want to start sending data after we're connected, so we put the `setInterval` inside `onConnected`.

5. And we're done. The final code looks like this, and should be entered on the right-hand side of the IDE and then uploaded by clicking the *Upload* button:

```
var WIFI_NAME = "";
var WIFI_KEY = "";
var wifi;

function onInit() {
  USB.setConsole(true);
  Serial1.setup(115200, { tx: B6, rx : B7 });
  wifi = require("ESP8266WiFi_0v25").connect(Serial1, function(err) {
    if (err) throw err;
    console.log("Connecting to WiFi");
    wifi.connect(WIFI_NAME, WIFI_KEY, function(err) {
      if (err) throw err;
      onConnected();
    });
  });
}

function onConnected() {
  console.log("Connected");
  setInterval(sendDweet, 10000);
}


function sendDweet() {
  var str = E.getTemperature().toFixed(2);
  console.log("Sending "+str);
  var url = "http://dweet.io/dweet/for/espruino_tmp?temp="+str;
  require("http").get(url, function(res) { });
}
```

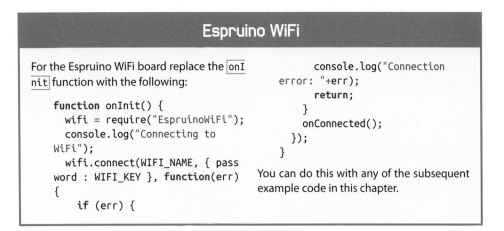

1. Finally, upload the code, type `save()` on the lefthand side, and you're done. Espruino will restart, start connecting, and will then automatically start sending data to dweet.

Experiment 31: Getting Data from the Internet

So now that we've managed to send data to a service on the internet, it would be nice to be able to get data from the internet and do something based on it.

Probably the simplest way to do this is to request a web page from the internet. We'll use dweet.io again, but this time in the other direction. As before, you might want to change the name we've used here (`espruino_led`) to something else.

We'll use exactly the same `onInit` function as before, but will change it so that we request the page *https://dweet.io/get/latest/dweet/for/espruino_led* from dweet.io (getting the state of the `espruino_led` device). This is a bit like the follow page *http://dweet.io/follow/ espruino_led*, but it's designed to be machine readable (and doesn't auto-update).

1. First, we need some data. Navigate to *http://dweet.io/dweet/for/espruino_led?on=0* in your web browser. This will send some data to dweet.io, setting the value `on` to `0`.

   Your browser will display something like this:

```
{
  "this":"succeeded",
  "by":"dweeting",
  "the":"dweet",
  "with":{
    "thing":"espruino_led",
    "created":"2016-06-24T09:22:47.177Z",
    "content":{"on":0},
    "transaction":"a296e38e-906d-47e3-beb7-26e41b836393"
```

```
      }
    }
```

2. Now enter the following code on the righthand side of the Web IDE:

```
    var WIFI_NAME = "";
    var WIFI_KEY = "";
    var wifi;

    function onInit() {
      USB.setConsole(true);
      Serial1.setup(115200, { tx: B6, rx : B7 });
      wifi = require("ESP8266WiFi_0v25").connect(Serial1, function(err) {
        if (err) throw err;
        console.log("Connecting to WiFi");
        wifi.connect(WIFI_NAME, WIFI_KEY, function(err) {
          if (err) throw err;
          onConnected();
        });
      });
    }

    function onConnected() {
      console.log("Connected");
      setInterval(getDweet, 10000);
    }

    function onDweetData(data) {
      console.log("DWEET> "+data);
    }

    function getDweet() {
      var url = "http://dweet.io/get/latest/dweet/for/espruino_led";
      require("http").get(url, function(res) {
        var data = "";
        res.on('data', function(d) { data+=d; });
        res.on('close', function() {
          onDweetData(data);
        });
      });
    }
```

3. Upload, and type `onInit()` on the lefthand side.

   This will retrieve data from dweet.io every 10 seconds as a string. It should look a bit like this:

```
    DWEET> {"this":"succeeded","by":"getting","the":"dweets","with":
    [{"thing":"espruino_led","created":"2016-06-24T09:22:47.177Z","content":
    {"on":0}}]}
    DWEET> {"this":"succeeded","by":"getting","the":"dweets","with":
    [{"thing":"espruino_led","created":"2016-06-24T09:22:47.177Z","content":
    {"on":0}}]}
```

4. Now visit *http://dweet.io/dweet/for/espruino_led?on=1*, and you should see that the value of `on` in the result changes the next time a request is made.

The data you're receiving is formatted in JSON. While we could search the string for the text `"on":`, that's not a very robust way of handling the data.

Instead, because Espruino uses JavaScript, it has a JSON parser built into it.

5. To parse the data properly we can just change the `onDweetData` function to the following:

```
function onDweetData(data) {
  var json = JSON.parse(data);
  console.log("DWEET> ", json);
}
```

(You can just enter this on the lefthand side of the IDE, rather than having to reconnect to WiFi.) The output will now be:

```
DWEET>  {
  "this": "succeeded",
  "by": "getting",
  "the": "dweets",
  "with": [
    {
      "thing": "espruino_led",
      "created": "2016-06-24T09:22:47.177Z",
      "content": { "on": 1 }
    }
  ]
}
```

While this is the same data, Espruino has read the data into the variable `json`. When `console.log` prints it, it's able to indent everything correctly. If the string wasn't valid JSON, `JSON.parse` would throw an exception.

Finally, we can decode this data, but we can't guarantee that we'll always get the right response. We could write something like this that checks that every bit of data we expect is there:

```
function onDweetData(data) {
  json = JSON.parse(data);
  console.log("DWEET> ", json);
  if (json && json.this=="succeeded" &&
      json.with && json.with[0] &&
      json.with[0].content &&
      json.with[0].content.on !== undefined) {
    var d = json.with[0].content.on;
    digitalWrite([LED2,LED1], d);
  } else
    console.log("Error decoding dweet");
}
```

But it's much cleaner just to use exceptions.

6. Enter the following code:

```
function onDweetData(data) {
  try {
    json = JSON.parse(data);
    console.log("DWEET> ", json);
    var d = json.with[0].content.on;
    digitalWrite([LED2,LED1], d);
  } catch (e) {
    console.log(e.toString());
  }
}
```

If there is an error in the code, it will be *caught* and printed, without causing problems for any of the code that called it.

So now, the red LED should have lit up, because on dweet.io, on was set to 1.

7. Go to *http://dweet.io/dweet/for/espruino_led?on=2* in your browser.

Within 10 seconds, the LED should have changed to green. You can load the URL again with different numbers for on between 0 and 3 to set a new state for the LEDs.

If you had another Espruino board connected up with WiFi you'd now be able to use it to control the first board by using the temperature experiment's code and modifying the URL from http://dweet.io/dweet/for/espruino_tmp *to* http://dweet.io/dweet/for/espruino_led.

Isn't There a Better Way?

Polling for data (where code repeatedly requests a web page, looking for new information) is often used by IoT devices. It's nice and easy, but it isn't very fast. When you change some information you have to wait for the device to poll (check the web page) again before it updates. There's a balance to be struck: if you poll often then the device will update more quickly, but you'll use up more data (you might be on a mobile network, where you have bandwidth limits) and if you have many devices, your servers may become overwhelmed.

Some servers offer the ability to keep an HTTP connection open; *http://dweet.io/listen/for/* *dweets/from/espruino_led* on dweet.io, for example, uses "chunked" HTTP to send an infinite web page that gets longer whenever the value changes state. The URL won't work on your web browser because the browser expects the web page to end at some point (and it never does), but it is possible to handle it with Espruino.

Infinite web pages aren't well supported, but there is a standard called WebSockets that allows a computer to open an HTTP connection and send and receive chunks of data in a well-defined way. To see how to implement this on Espruino, check out the Espruino website (*http://www.espruino.com/ws*).

Of course, there's no need to use HTTP (which is generally meant for web pages) at all. You could open a plain TCP/IP connection to a server, or could even use a well-defined protocol such as MQTT (*http://www.espruino.com/MQTT*).

These methods all require a connection to be made from the device out to some other server. The main reason for this is that most computers on a local network will be behind some kind of router. They won't have an IP address that can be used to uniquely access them from the internet.

However, if you have configured your router to allow incoming data to be transferred to your device, or you're only contacting the device on your local network, you can set your device up as a server. That means that you can access it directly, and can contact it only when you want it to do something.

Experiment 32: Creating a Server

As mentioned, sometimes it makes sense not to poll, but to contact the device only when you want data from it or need to tell it to do something. This is where creating a web server can come in handy.

First, we'll use the same initialization code as before, but in `onConnected` we'll create a web server on port 80 (the standard HTTP port) with `require("http").createServer`. When a web page is requested `onPageRequest` will be called, which will be responsible for providing the web page.

1. Enter the following code on the righthand side of the IDE, hit *Upload*, and then type `onInit()` on the lefthand side:

```
var WIFI_NAME = "";
var WIFI_KEY = "";
var wifi;

function onInit() {
  USB.setConsole(true);
  Serial1.setup(115200, { tx: B6, rx : B7 });
  wifi = require("ESP8266WiFi_0v25").connect(Serial1, function(err) {
    if (err) throw err;
    console.log("Connecting to WiFi");
    wifi.connect(WIFI_NAME, WIFI_KEY, function(err) {
      if (err) throw err;
      onConnected();
    });
  });
}

function onConnected() {
  console.log("Connected");
  require("http").createServer(onPageRequest).listen(80);
  wifi.getIP(function(err,ip) {
    console.log("Your IP address is http://"+ip);
  });
}
```

```
function onPageRequest(req, res) {
  console.log("Serving "+req.url);
  res.writeHead(200);
  res.end("Hello World");
}
```

After a few seconds you should see something like the following:

```
Connecting to WiFi
Connected
Your IP address is http://192.168.1.156
```

2. Go to that web address in your web browser, and you should see a simple web page served up directly from the Espruino board:

3. Look back at the Web IDE; it'll probably say:

```
Serving /
Serving /favicon.ico
```

This means there have been two requests: one for `/favicon.ico` (the icon for the website) and one for `/`, the main index page.

To do things properly, we should really serve up a *not found* web page for web pages we don't understand. After all, *Hello World* isn't an icon.

4. Change `onPageRequest` to the following:

```
function onPageRequest(req, res) {
  console.log("Serving "+req.url);
  var r = url.parse(req.url);
  if (r.pathname == "/") {
    res.writeHead(200);
    res.end("Hello World");
  } else {
    res.writeHead(404);
```

```
        res.end("404 - Not Found!");
    }
}
```

 We're using url.parse *here because the URL string could contain extra arguments. For example, when we sent data to dweet.io we used* ?temp=... *. In this case we'd still want to serve up the page for* / *, even though* url *might be* /?temp... *.*

5. Refresh the page a few times. You'll see that /favicon.ico was only requested the first time. Since the Espruino told the web browser it doesn't exist, the web browser has stopped trying to request it.

6. If you now go to *http://IP_ADDRESS/foo*, you'll get a 404 - Not Found! message, because that page doesn't exist.

   If you want to serve up something special for it, you can just add another line to the if statement.

   So now that we know how to serve up pages, let's create a page to show the current temperature.

7. Add homePage and change onPageRequest to the following:

```
var homePage = '<html><body>'+
 '<h1>My Espruino</h1>'+
 '<a href="/getTemp">Temperature</a>'+
 '</body></html>';

function onPageRequest(req, res) {
  console.log("Serving "+req.url);
  var r = url.parse(req.url, true);
  if (r.pathname == "/") {
    res.writeHead(200, {"Content-Type": "text/html"});
    res.end(homePage);
  } else if (r.pathname == "/getTemp") {
    res.writeHead(200, {"Content-Type": "text/html"});
    res.end('<html><head>'+
      '<meta http-equiv="refresh" content="2">'+
      '</head><body>'+E.getTemperature().toFixed(2)+
      '</body></html>');
  } else {
    res.writeHead(404);
    res.end("404 - Not Found!");
  }
}
```

This code doesn't just serve up plain text, it serves up HTML (stored in the `home Page` variable), with a hyperlink. If you go back to the main IP address in your web browser you should now see something like this:

Why Store the Page in a Variable?

In Espruino, it's more efficient to store `home Page` in a variable. If you store it as a string in a function then it will take up space in that function as part of the code, but when the function is executed the string will also be duplicated.

However, we're not putting the Temperature page in a separate variable because it has to change all of the time.

8. Click *Temperature*, and it will load a web page that automatically refreshes every two seconds (because of the *meta refresh* HTML tag), and shows you the current temperature of Espruino.

Controlling Things

So what if we wanted to control something on our Espruino? The easiest way is to supply arguments in the URL. The `url.parse` command we used earlier has already decoded them, so it's just as easy as checking `r.query` for the arguments we want.

1. Modify `homePage` and the `if (r.pathname == "/") {` part of the `if` statement, so `onPageRequest` looks like the following:

```
var homePage = '<html><body>'+
 '<h1>My Espruino</h1>'+
 '<a href="/getTemp">Temperature</a><br>'+
 '<a href="/?led=1">LED ON</a><br>'+
 '<a href="/?led=0">LED OFF</a><br>'+
'</body></html>';

function onPageRequest(req, res) {
  console.log("Serving "+req.url);
  var r = url.parse(req.url, true);
  if (r.pathname == "/") {
    // if an argument is given, set the LED state
    if (r.query && r.query.led)
      digitalWrite(LED1, r.query.led);
    // serve up the page
    res.writeHead(200, {"Content-Type": "text/html"});
    res.end(homePage);
  } else if (r.pathname == "/getTemp") {
    res.writeHead(200, {"Content-Type": "text/html"});
    res.end('<html><head>'+
      '<meta http-equiv="refresh" content="2">'+
      '</head><body>'+E.getTemperature().toFixed(2)+
      '</body></html>');
  } else {
    res.writeHead(404);
    res.end("404 - Not Found!");
  }
}
```

Request the page again. You should now see the following:

2. Click on *LED ON* or *LED OFF*, and the LED will turn on and off!

   Finally, we mentioned a few experiments back that if you wanted to change state, it was much better to use an HTTP POST request, not a GET (which is the default if you just load a URL). You can do this using forms.

3. Tweak homePage and onPageRequest to the following:

```
var homePage = '<html><body>'+
 '<h1>My Espruino</h1>'+
 '<a href="/getTemp">Temperature</a><br>'+
 '<form action="/?led=1" method="post">'+
 '<input type="submit" value="On"/></form>'+
 '<form action="/?led=0" method="post">'+
 '<input type="submit" value="Off"/></form>'+
'</body></html>';

function onPageRequest(req, res) {
  console.log("Serving "+req.url, req);
  var r = url.parse(req.url, true);
  if (r.pathname == "/") {
    // if an argument is given, set the LED state
    if (req.method=="POST" && r.query && r.query.led)
      digitalWrite(LED1, r.query.led);
    // serve up the page
    res.writeHead(200, {"Content-Type": "text/html"});
    res.end(homePage);
  } else if (r.pathname == "/getTemp") {
    res.writeHead(200, {"Content-Type": "text/html"});
    res.end('<html><head>'+
      '<meta http-equiv="refresh" content="2">'+
      '</head><body>'+E.getTemperature().toFixed(2)+
      '</body></html>');
  } else {
    res.writeHead(404);
    res.end("404 - Not Found!");
  }
}
```

onPageRequest now checks req.method to see if it is a POST before performing any actions, and two <form> elements load a specific URL when each button is pressed:

 You can transfer all kinds of data using forms, but the information is generally not transferred in the URL, but in the body of the HTTP request, which makes it slightly more painful to extract. See the Espruino website (http://www.espruino.com/Internet) for more information.

You can, however, use JavaScript and `XMLHttpRequest` on the webpage to POST data back as easily parseable JSON, much as we did in the last chapter.

And that's it! You're well on your way to creating your own amazing Internet of Things devices!

Other Connection Types

While we've covered ESP8266 WiFi in this chapter, Espruino also supports other ways of connecting to the internet.

For instance, you can use Ethernet using a WIZnet Ethernet module (*http://www.espruino.com/WIZnet*), or can connect over GSM/GPRS using SIMCom SIM800 or SIM900 modules (*http://www.espruino.com/SIM900*).

You can even install Espruino on the ESP8266 WiFi module (*http://www.espruino.com/EspruinoESP8266*) itself!

ES6 Template Literals

Towards the end of this chapter, defining the `homePage` *variable started looking quite ugly as we added more and more text. Espruino supports ES6's Template Literals, which allow you to create multiline strings using the backtick character instead of normal quotes.*

For example, consider:

```
var homePage = '<html><body>'+
'<h1>My Espruino</h1>'+
'<a href="/getTemp">Temperature</a><br>'+
'<form action="/?led=1" method="post">'+
'<input type="submit" value="On"/></form>'+
'<form action="/?led=0" method="post">'+
'<input type="submit" value="Off"/></form>'+
'</body></html>';
```

This could just be written as:

```
var homePage =
`<html><body>
<h1>My Espruino</h1>
<a href="/getTemp">Temperature</a><br>
<form action="/?led=1" method="post">
<input type="submit" value="On"/></form>
<form action="/?led=0" method="post">
<input type="submit" value="Off"/></form>
</body></html>`;
```

Bluetooth Low Energy

So far we've looked at all kinds of wireless communication, but we've been missing one very common type: Bluetooth.

Bluetooth works on the 2.4GHz radio band like WiFi does, but it's designed for lower power usage, lower bandwidth, and shorter communication distance than WiFi. Probably the most common use of Bluetooth is now in wireless headphones and speakers.

However, there is a problem. Bluetooth is more power efficient than WiFi, but it's not *that* efficient, and it's still pretty complicated to implement. It usually requires one microcontroller to handle Bluetooth, and another to do everything else (like interacting with the device's user).

Bluetooth Low Energy was created to try to fill that gap. It is very power efficient, and is simple enough that it can be implemented on a microcontroller alongside software that handles other things such as a device's user interface.

Probably the most exciting thing about Bluetooth Low Energy is that it is now built into pretty much every new PC, laptop, tablet, and phone. Unlike the 433Mhz radio we looked at, you don't need any special hardware to *bridge* between the wireless device and your phone or PC.

As you could with the ESP8266 and WiFi you can buy extra boards that can be added to a microcontroller to give it Bluetooth LE capability. The HM-10 module is a very common module, or there are many others including the Bluefruit LE series of boards from Adafruit.

Using Normal Bluetooth

If you want to use *normal* non–low-energy Bluetooth with Espruino, you can still do that. The original Espruino boards have an area where an HC-05, HC-06, or HM-10 Bluetooth module can be soldered on (Figure 16-1).

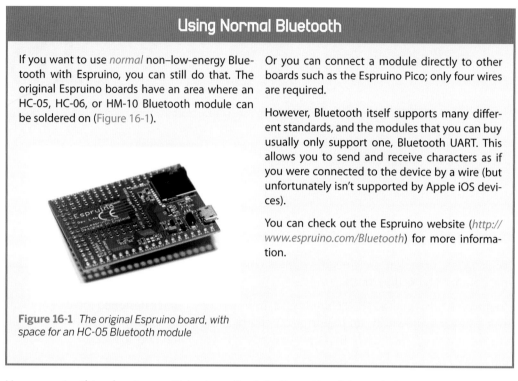

Figure 16-1 *The original Espruino board, with space for an HC-05 Bluetooth module*

Or you can connect a module directly to other boards such as the Espruino Pico; only four wires are required.

However, Bluetooth itself supports many different standards, and the modules that you can buy usually only support one, Bluetooth UART. This allows you to send and receive characters as if you were connected to the device by a wire (but unfortunately isn't supported by Apple iOS devices).

You can check out the Espruino website (*http://www.espruino.com/Bluetooth*) for more information.

However, in this chapter we'll look at Puck.js (*http://puck-js.com*) (Figure 16-2). This is another Espruino device, but it has Bluetooth Low Energy built in (it doesn't need an add-on board).

Inside, Puck.js contains a chip called the Nordic nRF52832. This is an ARM microcontroller (like the one in the Espruino Pico), but it also contains a Bluetooth radio as well. Because everything is contained in one chip it's very power efficient, and can last for around a year on the built-in CR2032 battery.

Figure 16-2 *Puck.js*

So, How Does Bluetooth Low Energy Work?

Bluetooth Low Energy has two main modes: the first is called *advertising*. This is where a Bluetooth LE device broadcasts information at set intervals. There's no two-way communication, it is just the device sending out information to anything within range.

Normally the data that is sent out contains the device's name and information about the functionality that can be controlled wirelessly (called *services*). If you're ever asked to connect to a Bluetooth LE device on your phone, the names of devices you'll see on the screen are almost certainly the result of the advertising messages sent by the devices.

Advertising can be used in other ways, too. Beacons (common examples are iBeacon, Eddystone, and AltBeacon) are devices that broadcast a unique ID or URL (web address) to any device in range. This can be used to help locate you indoors, or can be used to provide location-specific information: in a museum your phone might provide a link to information on the exhibit you're standing by.

The second mode is when one device like a phone or PC (referred to in Bluetooth jargon as a *central*) connects to another device (the *peripheral*).

When this happens, the central gets access to a list of *characteristics* on the peripheral. Characteristics are just bits of data that can be changed. They can be read, written, or the central can request to be notified whenever one of them changes.

For example, a Bluetooth Low Energy light bulb might have a characteristic for the brightness of the light bulb, and another for whether the bulb should be on or off. When a central connects to the bulb and writes a `0` to the *on-off* characteristic, it turns the light bulb off, or a `1` turns it on.

A Bluetooth LE button might have a characteristic called *pressed*. A central device connecting to it wouldn't be able to write data to the characteristic, but it could read its state, and could request to be notified when it changed (so it would know as soon as the button was pressed without having to constantly read the values).

In reality, the services and characteristics don't have human-readable names, they have UUIDs (Universally Unique Identifiers). The Bluetooth SIG (the standards body responsible for Bluetooth) keeps a list of IDs (*http://bit.ly/2qWNYL2*), but you can make your own.

The UUIDs provided by the Bluetooth SIG are 16 bits (so there can only ever be $2^{16} = 65536$ of them), but when you make your own you need to use 128-bit UUIDs of the form `abcdabcd-abcd-abcd-abcd-abcdabcdabcd`. There are so many possible combinations available in 128 bits ($2^{16} = 3.4 * 10\wedge38$) that the Bluetooth SIG suggest that you just randomly choose a number, and the chances of it being taken are so small you don't have to worry.

This may all seem a bit complex, but it does provide a very well-defined interface between hardware. You can load up an app like *nRF Connect* or *LightBlue LE* and list the characteristics of your shiny new Bluetooth LE device to see what it implements. A quick Google search for the UUIDs may even bring you to a website with some explanation showing you how to use it!

How Can We Use Bluetooth LE Ourselves?

Nordic (which makes the microcontroller that Puck.js uses) has come up with a service and some characteristics to handle transmission and reception of streams of data; they call this *Nordic UART* (Figure 16-3).

This service has the UUID `6e400001-b5a3-f393-e0a9-e50e24dcca9e`.

By creating the service and characteristics with the same UUIDs a device can signal to other apps that it can be communicated with just by sending characters.

There are a load of apps around that will communicate with these UUIDs— two of the better ones are *Adafruit Bluefruit LE* app and *nRF UART*.

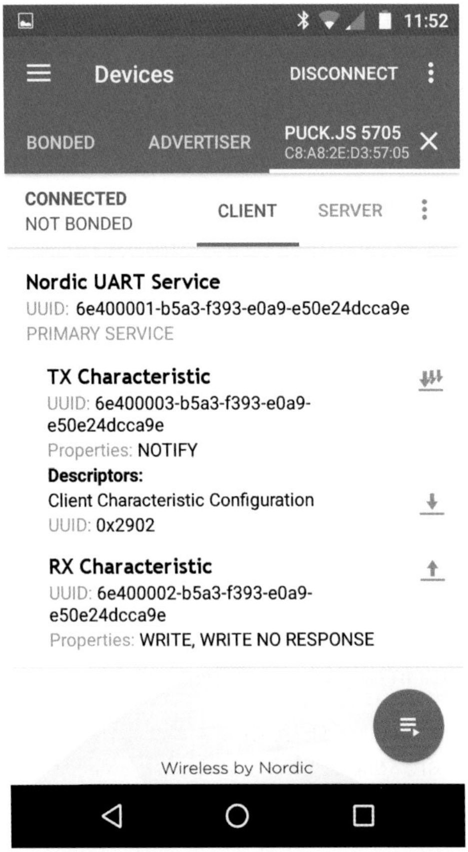

Figure 16-3 *The nRF Connect app showing the Nordic UART service reported by Puck.js*

By default Puck.js implements the Nordic UART service shown in Figure 16-3 (you can customize the services and characteristics it has). This allows you to program and debug it completely wirelessly.

Web Bluetooth

Finally we come to Web Bluetooth, a standard for web browsers that allows you to access Bluetooth Low Energy devices straight from a web page.

Your initial thoughts might be of huge security issues, but Web Bluetooth has been designed from the ground up to be as secure as possible. Every time a web page wants access to a Bluetooth device it'll have to ask you with a pop-up window, in the same way a web page asks to access your webcam.

Web Bluetooth is great news for everyone, especially makers. Before, you'd have had to make a different app for each device: Android, iOS, Windows, Mac, and maybe even some other platforms.

With Web Bluetooth, you can make one single website (which you can easily keep up-to-date), and with it you can control your Bluetooth LE devices from any platform!

Experiment 33: Using Puck.js

You'll need:

- A Puck.js device

First you need to turn your Puck.js on. As it comes it has a small piece of plastic between the battery and the PCB to stop it from being powered on in transit.

1. Gently peel back the silicone case with your fingers and remove it.

2. Tip the PCB out of the black case.

3. Now use a blunt object to push the battery out of the battery holder from behind: a ballpoint pen or matchstick is great for this.

4. Remove the piece of plastic. It's clear and circular and can be quite hard to see!

5. Reinsert the battery. The red LED should flash on for just a fraction of a second.

6. Now you need to reassemble it. Put the Puck back in the case with the battery facing down and the silver Bluetooth module facing up. The text saying `www.puck-js.com` should be resting against the shelf in the case.

   When it's correct, Puck.js will sit flat in the case and there's a satisfying *click* sound when it's pressed.

   Now it's time to start controlling it.

7. Go to *https://www.puck-js.com/go*, where you'll find information about whether the device you're using supports Web Bluetooth, and how to set it up so that you can get a working Web IDE.

Web Bluetooth Support

When Web Bluetooth rolls out across all browsers, you'll simply be able to visit https://espruino.com/ide in your browser and start writing code—exactly as you did with the Chrome-based Web IDE.

Until then you may have to do a little bit of setting up on some platforms. Instructions can be found at the Puck.js website (https://www.puck-js.com/go).

---

8. Now that you've got a working IDE you're ready to go. You can follow the instructions in Chapter 3 to try controlling it; for instance, after connecting you can just

type `LED.set()` to light an LED, or `E.getTemperature()` to get the current temperature.

Controlling Puck.js from an App

Because Puck.js uses the Nordic UART service, other apps can control it too.

1. On your Android or iOS phone, install the *Adafruit Bluefruit LE* app and run it.

2. Tap *Connect* next to the `Puck.js ABCD` device in the device list (on the left in iOS).

3. In the pop-up window on Android, tap *UART*, or on iOS, tap UART at the bottom of the screen (Figure 16-4).

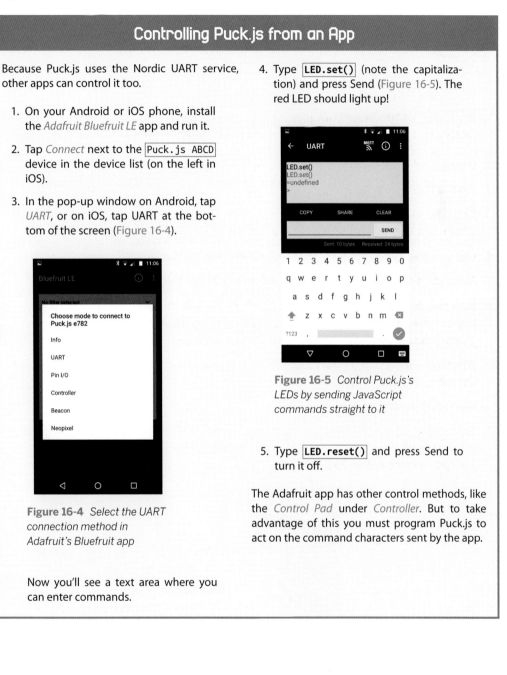

Figure 16-4 *Select the UART connection method in Adafruit's Bluefruit app*

Now you'll see a text area where you can enter commands.

4. Type `LED.set()` (note the capitalization) and press Send (Figure 16-5). The red LED should light up!

Figure 16-5 *Control Puck.js's LEDs by sending JavaScript commands straight to it*

5. Type `LED.reset()` and press Send to turn it off.

The Adafruit app has other control methods, like the *Control Pad* under *Controller*. But to take advantage of this you must program Puck.js to act on the command characters sent by the app.

Experiment 34: Making a Door Opening Counter

For this experiment we're going to use Puck.js's internal magnetometer to measure if a magnet is close or not, and will then count the number of times a door has been opened and transmit it wirelessly.

You'll need:

- A neodynium magnet
- Tack-it, Blu-tack, or other adhesive putty

Here are the steps:

1. First, add some of the adhestive putty to the magnet and stick it in the corner of the door, farthest from the hinge.

    If you can orient the magnet so that the magnetic field is facing vertically you'll get the best results. On disc magnets this means that the flat surface is parallel to the ground. Unless you have loosely fitting doors this may be tricky, but don't worry if not. The magnets are strong and Puck.js's magnetometer is relatively sensitive, so positioning the magnet flat to the door as shown here will work fine, too:

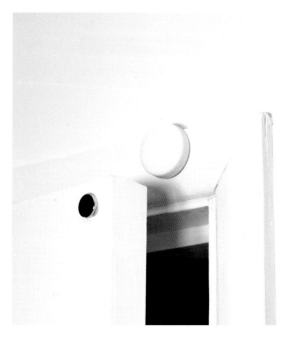

2. Now use the adhesive putty to attach the Puck.js to the door frame just above the corner of the door where the magnet is.

3. Connect to Puck.js with the Web IDE, and type `Puck.mag()`. This will return the magnetometer reading in three dimensions.

4. Try it with the door open and closed. You should see different readings depending on the two states:

```
// door open
>Puck.mag()
={ "x": -1, "y": -1265, "z": -665 }
>Puck.mag()
={ "x": -3, "y": -1262, "z": -665 }
>Puck.mag()
={ "x": -3, "y": -1268, "z": -651 }
// door closed
>Puck.mag()
={ "x": -539, "y": -2031, "z": -1333 }
>Puck.mag()
={ "x": -549, "y": -2066, "z": -1352 }
// door open
>Puck.mag()
={ "x": -3, "y": -1244, "z": -642 }
```

So now we just need to figure out which is which.

5. Open the door fully, enter the following code on the righthand side of the IDE, and click *Upload*:

```
// Magnetism measured when nothing around
var zeroMag = Puck.mag();

// Called when new magnetic field information is found
function onMag(xyz) {
  // Work out the distance from zero
  var x = xyz.x - zeroMag.x;
  var y = xyz.y - zeroMag.y;
  var z = xyz.z - zeroMag.z;
  // Work out the magnitude of the field
  var d = Math.sqrt(x*x + y*y + z*z);
  // Print it and light the light if the door is closed
  console.log(d);
  LED2.write(d>500);
}

// Set callback when magnetic field info is found
Puck.on('mag', onMag);
// Turn on magnetometer
Puck.magOn();
```

6. Now try closing the door. The green LED should light up.

You should also see a series of readings output on the console:

```
29.22327839240
45.61797891182
46.62617290749
1222.32156161952
2571.24269566293
71.58910531638
54.53439281774
54.21254467371
52.92447448959
43.66921112179
47.67598976424
```

The largest readings are from when the door is closed. If your LED doesn't light up, you may need to change the value 500 in the preceding code to something smaller.

*The preceding code is comparing the magnetometer reading with the reading taken at the time the code was uploaded. Make sure the door is always open when you upload code, or Puck.js will get the wrong **zero** reading.*

Ok, now we can detect when the door is open or closed. Let's detect when it changes, and then we can count the number of openings.

7. Change the code to the following by adding wasDoorOpen, doorOpenings, and doorOpened, and by changing onMag:

```
// Magnetism measured when nothing around
var zeroMag = Puck.mag();
var wasDoorOpen = false;
var doorOpenings = 0;

function doorOpened() {
  doorOpenings++;
}

// Called when new magnetic field information is found
function onMag(xyz) {
  // Work out the distance from zero
  var x = xyz.x - zeroMag.x;
  var y = xyz.y - zeroMag.y;
  var z = xyz.z - zeroMag.z;
  // Work out the magnitude of the field
  var d = Math.sqrt(x*x + y*y + z*z);
  // Check door open state
  var isDoorOpen = d<500;
  if (isDoorOpen != wasDoorOpen) {
    if (isDoorOpen) {
      doorOpened();
```

```
        // Flash green LED for open
        digitalPulse(LED2,1,100);
    } else {
        // Flash red LED for close
        digitalPulse(LED1,1,100);
    }
    wasDoorOpen = isDoorOpen;
  }
}


// Set callback when magnetic field info is found
Puck.on('mag', onMag);
// Turn on magnetometer
Puck.magOn();
```

8. Upload the code again.

   Now, when you close the door the red LED should flash, and when you open it the green LED will flash. If you type `doorOpenings` on the console, it will display the number of times it thinks the door has opened and closed.

 By default the magnetometer only checks once every 1.5 seconds, so if you were able to open the door and close it within 1.5 seconds, at exactly the right time, you might be able to fool the sensor!

You can supply a value to `Puck.magOn()`, for example, `Puck.magOn(5)`, to increase the sample rate to five times a second, but this will also decrease battery life!

As is, we can use Puck.js like this. An app or Web Bluetooth page could connect to Puck.js, enter the characters `doorOpenings` and a newline, and Puck.js would output the number of door openings.

If you upload the following to an HTTPS-capable website like GitHub Pages (unfortunately JSFiddle won't work because of its use of frames) and then click on the big ? on the screen, the web page will report the number of times the door has opened.

To update the value, you can just click again:

```
<html>
 <head><title>Door Opening reader</title></head>
 <body>
  <script src="https://www.puck-js.com/puck.js"></script>
  <p style="text-align:center">
    Door opened<br/>
    <span id="result" style="font-size:200px;cursor:pointer">?</span><br/>
    times...
  </p>
```

```
<script>
  function onClick() {
    Puck.eval("doorOpenings", function(result) {
      document.getElementById("result").innerHTML = result;
    });
  }
  document.getElementById("result").addEventListener("click", onClick);
</script>
</body>
</html>
```

However, that's not ideal. It would be much better if Puck.js could just advertise the information, as then a connection doesn't have to be maintained, and other devices could just listen and be notified when it changed.

Experiment 35: Advertising Door Openings

To do this, we just need to use the `NRF.setAdvertising` function.

1. Change the `doorOpened` function to this:

```
function doorOpened() {
  doorOpenings++;
  NRF.setAdvertising({
    0xFFFF : [doorOpenings]
  });
}
```

2. Disconnect the Web IDE (Puck.js can't advertise while it is connected to another device).

   Now, when the door is opened or closed, Puck.js will advertise a single byte value representing the number of times the door has been opened to date (so if you open the door more than 255 times it will roll back around) under the service UUID `0xFFFF`.

 The UUID `0xFFFF` is used here for testing. As previously discussed, the 16-bit UUIDs are assigned by the Bluetooth SIG, so if you were creating a device that you were going to sell you should use the correct UUID for what you plan to advertise, or should use a totally random 128-bit UUID.

---

3. Install the *nRF Connect* app on your Android or iOS phone (the LightBlue app works too) and open it up.

   You should see a list of devices, of which `Puck.js ABCD` will be one.

4. Tap on the device (but not on the *Connect* button).

It should open out and show you something like this, where you can see $\boxed{\texttt{0xFFFF}}$ and a hex value that represents the number of times the door has been opened:

If you were willing to write an Android or iOS app that could run in the background on your phone and listen for the advertising packets from Puck.js, the current solution of setting advertising data would be perfect.

But let's assume we don't want to or can't make an app. What could we do? Well, this is where beacons come in. We mentioned earlier that beacons could transmit information like URLs. We could advertise a URL that contained the number of times the door has opened.

For this, we're going to use Eddystone.

Experiment 36: Receiving Door Openings with Eddystone

1. First, create a web page using HTTPS (maybe via GitHub Pages again) with the following HTML on it.

   If you don't want to create your own page then I've put one online already. It's at *https://www.espruino.com/dooropen.html*.

```html
<html>
 <head><title>Door Openings</title></head>
 <body>
  <p style="text-align:center">
    Door opened<br/>
    <span id="result" style="font-size:200px">?</span><br/>
    times...
  </p>
  <script>
   if (window.location.hash)
      document.getElementById("result").innerHTML = window.loca
tion.hash.substr(1);
  </script>
 </body>
</html>
```

This web page is really simple. It displays *Door opened ? times,* and if you put a #️ followed by some text after the URL, it displays that text.

2. Enter *https://www.espruino.com/dooropen.html#42* in your browser's address bar, and you should see this:

So now all we have to do is to get Puck.js to serve up that URL. Unfortunately, we can't do it directly as the URL is too long, so we need to use a URL shortener like *https://goo.gl* to shorten the URL. I've done this already with *https://goo.gl/D8sjLK*.

3. Now connect to Puck.js again and change doorOpened to:

```
function doorOpened() {
  doorOpenings++;
  require("ble_eddystone").advertise("goo.gl/D8sjLK#"+doorOpenings);
}
```

> *On some platforms there can still be problems with the Eddystone implementation stripping off information after the #̲ character. If this happens to you then you could either use your own short URL, or could generate a different **goo.gl** URL for each different value you want to report.*

As this uses a library, you'll have to change the code on the righthand side of the IDE and re-upload rather than using the lefthand side of the IDE (make sure you do it with the door open).

4. Now disconnect from the IDE and open and close the door again to force Espruino to change the advertising.

5. If you look again with the *nRF Connect* app you should see something like this:

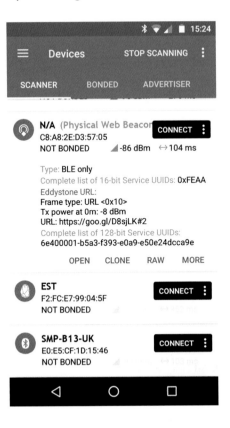

If you've got your phone set up properly to receive Eddystone, in a minute or two you should see a notification for the web page:

6. Click it, and it'll bring you to a web page with the information on the number of times the door was opened, without an app in sight!

Door opened

9

times...

Puck.js

*Puck.js is capable of so much more than this. It can even control other Bluetooth LE devices. Check out the Puck.js website (**https://www.espruino.com/Puck.js**) for more information.*

PART V

Putting It All Together

We've learned a lot so far, but the things we've made haven't really been able to produce anything of particularly high quality.

In these chapters we'll use what we've learned to create a machine that we might actually want to use!

XY Plotter

In Chapter 9 we looked at making a plotter using servo motors and chopsticks. While we could draw simple shapes, we couldn't draw anything very accurately.

Nearly all plotters actually use Cartesian coordinates. Therefore, one motor is responsible for moving the pen sideways, and one is responsible for moving it up and down. As most images are described this way on your computer, it makes it much easier to plot them on the plotter.

To get more precision, we're going to use off-the-shelf stepper motors this time. These particular motors are very easy to find; in fact, you may even find them in some consumer electronics!

Experiment 37: Making an XY Table

Most commercial machines use a toothed belt to make the stepper motors move the pen, but these can be expensive and hard to find.

Instead, we're going to use a mechanism that is very similar to an Etch-A-Sketch.

In an Etch-A-Sketch, each knob (for the x- and y-axes) is attached to a piece of wire that is looped around pulleys so that on each side of the Etch-A-Sketch there is some wire that travels in the same direction.

For instance, for our table the routing of the wire is shown in Figure 17-1. The y (vertical) axis is shown in red. When the pulley connected to the stepper motor (marked with a $\boxed{Y}$) is moved in a certain direction, the vertical inner pieces of wire on the left and right side will both move in the same direction.

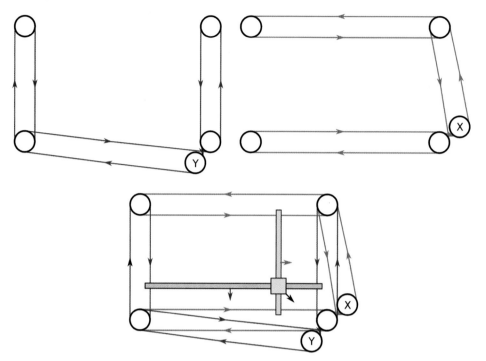

Figure 17-1 *The routing of wires around our pulleys for the x- and y-axes (note that each axis is on a separate set of pulleys, so it can move independently)*

A bar can then be attached to these pieces of wire, and the same can be done for the wire for the x-axis. We can now place a sled where the two bars overlap and can control the location of the sled in X and Y by moving the stepper motor.

You'll need:

- 1cm thick square sheet of plywood, at least 30cm×30cm
- Smooth aluminum bar, twice as long as your sheet of plywood is wide (so at least 60cm)
- A 30mm×30mm×30mm cube of solid wood or plastic
- 10× 40mm pulleys (the ones I used had a 3mm hole in the middle)
- 4x circular nails that will fit inside the pulleys
- 5 meters of fishing line
- 2x springs (springs from ballpoint pens will do!)
- 2x small geared stepper motors
- A pen (thin fiber-tipped pens work well as they don't need a lot of pressure to work well)

Follow these steps:

1. First, you need to draw a line along the top and left sides of your piece of wood, 2cm from the edge.

2. Then, do the same on the bottom and right sides of the wood. It should look like this:

3. Drill four holes where the lines you drew cross over (where the red Xs are lined up). The holes should be slightly smaller than the diameter of your nails so the nails are guided in when you hammer them in. If you have a drill press, it would be perfect to use for this.

4. Now put two pulleys on each one of your four nails, and hammer them into the holes you made, leaving them slack enough that your pulleys can turn easily:

The end of the nail will almost certainly come out of the bottom of the piece of wood, so be careful you don't nail your wood to the desk! You can cut the end of the nail off, or you could hammer another bit of wood onto the bottom and use it as a foot.

We've got the start of our table, so now we need to fit the two stepper motors for the X and Y axes. This is a little tricky as we want to position the pulleys so that the inner wire is pushed against the corner pulley while the diagonal wire that goes between the two corners *doesn't* touch it:

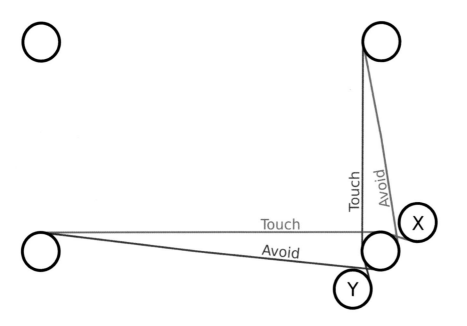

5. Draw the diagonal lines from the pulleys onto your sheet of wood. You can then place the pulleys on to the wood so that they are positioned correctly, and can draw around them.

6. Once that's done, fit the pulleys to your stepper motor.

   The pulleys we've used are made of very flexible plastic and it was enough to drill a 4.5mm hole in the pulley and just push it onto the stepper motors:

7. Now you need to fit your y-axis stepper motor (the one at the bottom) so that the pulley is in the position that you drew. You'll need to be careful here as the shaft of the stepper motor is off-center!

   I used a 30mm hole saw for this; however, you could cut a hole out of the wood however you feel comfortable. The hole doesn't have to be circular, it just has to fit the stepper motor while leaving somewhere to screw it down, so a simple slot in the wood is fine for this.

8. Finally, add the x-axis stepper motor in the same way, but this time you need to align the pulley on it with the upper pulleys rather than the lower ones.

This means you'll probably want to add a little offcut of wood (or a few washers) under each side of the motor to step it away from the plate.

Now it's time to route the two sets of wires:

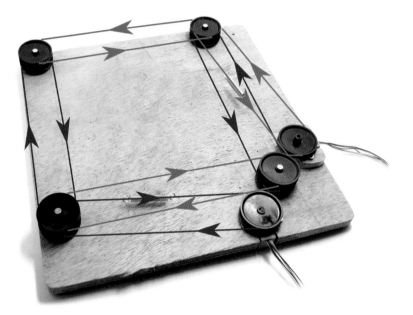

9. Route the bottom layer (the y-axis) first. Take a length of fishing wire and loop it around the pulleys following the direction of the red arrows in the diagram. You need to loop around the stepper motor twice.

10. Now that you've got an idea of the length, leave 6 inches of extra wire free and cut the fishing wire to length.

11. Tie one end of the fishing line to the spring.

12. Put the spring where it's placed in the preceding image (this is the longest stretch of space where it won't bump into anything).

13. Now route the wire around in the same way as before, and loop it through the other end of the spring.

14. Stretch the spring a little, tie the fishing line tight, and trim the excess.

    You should now have a working x-axis. If you turn the stepper motor you'll be able to feel the fishing line moving all the way around. The two innermost stretches of line opposite each other should be moving the same way.

15. Add the upper layer of fishing line for the x-axis, following the arrows and lines shown in red.

    Now that we have two working axes, the next step is to create two crossbars to go between them.

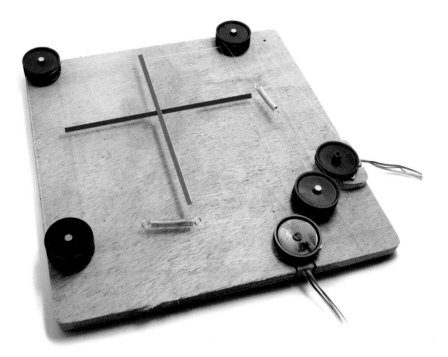

16. Measure between the two lengths of fishing line where shown for the red (Y) crossbar, add 1cm to the length, and cut the aluminum bar.

17. Cut a small groove in the aluminum bar 0.5mm from each end. Make sure both grooves are parallel as these will allow the bar to sit down flat on the fishing line. Do the same for the X crossbar (shown in green).

18. Place the two bits of bar with the grooves over the fishing rod so they are at right angles to each other.

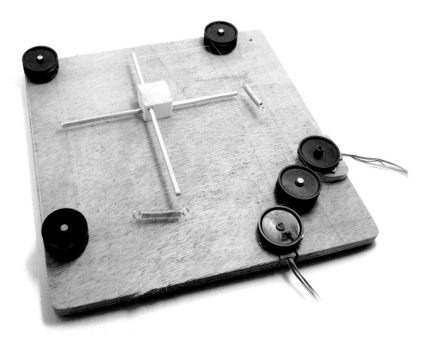

Now we need to make our sled.

19. Place the cube of wood/plastic by the side of the two crossbars and mark off the height of each.

20. Carefully drill a hole in one side of the cube for each rod, with the second hole perpendicular to the first. The holes should be big enough that the rods can move easily through it, but not so big that they wobble around.

21. Measure the diameter of your chosen pen and find a drill that will make a hole that will hold your pen snugly. Drill a hole in the sled at right angles to the other two holes. Make sure it is in a part of the sled that won't collide with the two holes you've already drilled.

    Now it's time to assemble the crossbars and sled. It might be tempting to grease the crossbars to make the sled slide more easily, but I'd advise against this. Normal grease tends to make it harder for the sled to slide, so if you want to grease the bars, use only a *dry* grease like teflon spray.

22. Rotate the stepper motors so the two springs are in the middle of their available travel.

23. Put the crossbars through the sled, and place them back over the fishing line.

24. Now glue them onto the fishing line. Superglue (cyanoacrylate) works well for this (especially if you have a *kicker* spray), but epoxy is good, too:

And finally we have a working XY table. Once the glue is dry, you can turn the Y stepper motor and the sled will move up and down, and you can turn the X stepper motor to move the sled left and right.

Experiment 38: Controlling the XY Table

You'll need:

- An Espruino WiFi or Pico board

- A breadboard

- 12 male-to-female jumper wires

- 2x ULN2003 driver boards (these usually come with your stepper motors)

Follow these steps:

1. Push the Espruino board into the breadboard so that the USB connector is as far over to the left as it will go.

2. Label the two motor driver boards, one as X and one as Y. These will be for the two stepper motors.

3. Connect the jumper wires between the breadboard and the motor driver boards as shown in the following table:

Motor driver	Espruino connection	Motor driver	Espruino connection
Driver X, -	GND	Driver Y, -	GND
Driver X, +	V_OUT	Driver Y, +	V_OUT
Driver X, IN1	B3	Driver Y, IN1	B10
Driver X, IN2	B4	Driver Y, IN1	B13
Driver X, IN3	B5	Driver Y, IN2	B14
Driver X, IN4	A6	Driver Y, IN3	B15

Once assembled, you should have something that looks like this:

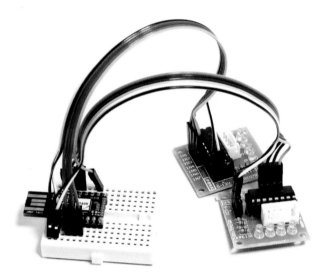

4. Connect the stepper motors to the driver boards, with the X stepper going to the X driver board. If your stepper motors and drivers came together there should be a keyed plug and socket to ensure that you connect them correctly.

5. Plug the Espruino board into your USB port, and connect with the Web IDE.

Now we'll interface to the stepper motors. While we controlled the motors directly in the first chapter, this time we'll use Espruino's `StepperMotor` library to make life easier.

6. Enter the following code on the righthand side of the IDE:

```
var StepperMotor = require("StepperMotor");
var motorx = new StepperMotor({
  pins:[B3,B4,B5,A6]
});
var motory = new StepperMotor({
  pins:[B10,B13,B14,B15]
});
```

7. Click *Upload*.

Nothing will happen immediately, because while we initialized the steppers we didn't tell them to do anything.

8. Type `motorx.moveTo(100)` in the lefthand side and press `Enter`.

The sled should now move. This will move to an absolute position, so typing `motorx.moveTo(100)` again won't do anything.

9. To move the sled back to its original position, type `motorx.moveTo(0)`.

10. Try `motory` and check if it works with `motory.moveTo(100)`.

You can also move to a position and then move back by chaining commands using callbacks.

11. Enter the following:

```
motorx.moveTo(100, 1000, function() {
  motory.moveTo(100, 1000, function() {
    motorx.moveTo(0, 1000, function() {
      motory.moveTo(0, 1000, function() {
        console.log("Done!");
      });
    });
  });
});
```

This will draw a square, by moving 100 steps in X, then Y, then backwards in X and then backwards in Y again. The `1000` in the function call specified that it should move 100 steps over the course of 1000ms (1 second).

This is moving much more accurately than the plotter we made before—the stepper motors are more precise, the wires are tigther, and the movement is linear throughout the whole area.

Now what if we want to be able to move diagonally? If we moved just one motor and then the other then we would end up with two lines at right angles.

To move diagonally we need to start moving both motors at the same time, and we need to set them so they both finish at the same time.

As we discussed in Chapter 5, there's a limit to how fast we can move the stepper motors, so we need to work out the distance we have to travel and from that the time we need to take to move each of the two motors.

Enter the following code on the righthand side of the IDE and upload again:

```
function moveTo(x,y,callback) {
    // Work out the distance in X and Y
    var dx = x - motorx.getPosition();
    var dy = y - motory.getPosition();
    // Work out the diagonal distance with pythagoras
    var d = Math.sqrt(dx*dx + dy*dy);
    // Work out how much time we've got to move
    var time = d * 1000 / motorx.stepsPerSec;
    // Set both motors moving
    motorx.moveTo(x, time);
    motory.moveTo(y, time, callback);
}
```

12. You can now type commands like moveTo(50, 100) and moveTo(-230, 25) and the sled will move diagonally to the correct location.

13. Let's try drawing a simple spiral pattern. Tape some paper down onto the wooden base, and put your pen in the sled.

14. Enter the following code:

```
function spiral(r,ang) {
  if (ang>=Math.PI*24) return;
  moveTo(Math.sin(ang)*r, Math.cos(ang)*r, function() {
    spiral(r+0.2, ang+Math.PI/40);
  });
}
spiral(0,0)
```

This will start drawing a spiral. You can change the values in the function to change how the spiral appears, but you should get something like this:

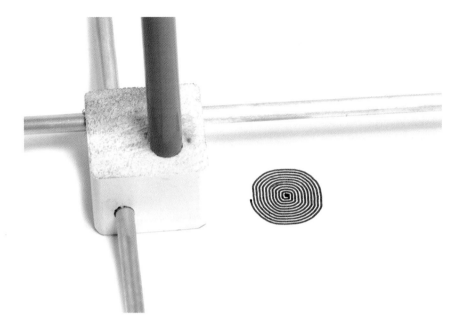

15. You won't be able to see most of the spiral immediately as the pen will be covering it. To move it out the way, you could unplug the Espruino from USB so the stepper motors power down, and then you can move the stepper motors' pulleys by hand. Either that or you can take the pen out and execute a command like `moveTo(0,500)` to move the sled away from the drawing.

    Now that we managed to draw something, let's try drawing something more interesting. To save us having to wait around for the plotter, we can experiment on the PC first.

16. Go to *https://jsfiddle.net*.

17. In the `HTML` area, type:

    ```
    <canvas id="canv" style="border:1px solid black"></canvas>
    ```

18. In the `JavaScript` area, type:

    ```
    var canvas = document.getElementById("canv");
    canvas.width = 500;
    canvas.height = 500;
    var ctx = canvas.getContext('2d');
    var midx = canvas.width/2, midy = canvas.height/2;
    var lastPos = [midx,midy];

    function moveTo(x,y,callback) {
      var pos = [x/4+midx, y/4+midy];
      ctx.beginPath();
      ctx.moveTo(lastPos[0], lastPos[1]);
    ```

```
    ctx.lineTo(pos[0], pos[1]);
    lastPos = pos;
    ctx.stroke();
    ctx.closePath();
    if (callback) setTimeout(callback,1);
  }
  // ----------------------------

  function spiral(r,ang) {
    if (ang>=Math.PI*24) return;
    moveTo(Math.sin(ang)*r, Math.cos(ang)*r, function() {
      spiral(r+0.2, ang+Math.PI/40);
    });
  }

  spiral(0,0);
```

19. Now click Run at the top. You'll see a spiral drawn:

You can now experiment with all kinds of different code to get interesting results, as long as you can make the image you want out of one long line!

You can even output 3D-looking shapes. For instance, try entering the following code after the `// ------------` comment:

```
  function sinrr(step, once) {
    // 'step' is going to keep increasing
    // make it 'scan' out in x and y
    var x = step % 100;
    var y = (step-x) / 100;
    if (y>=100) return;
    if (y&1) x = 100-x;
    // now center the coordinates on 0,0
    x -= 50;
```

```
    y -= 50;
    // Work out `r` - the radius
    //  - but add a bit to `r` to avoid a divide by 0 below
    var r = Math.sqrt(x*x + y*y) + 0.1;
    // Make 'z' a fun mathematical formula - (sin r)/r in this case
    var z = 100 * Math.sin(-r/2) / r;
    // now work out some 3D coordinates
    var a = 0.4; // rotation in 'y' axis
    var b = 0.5; // rotation in 'x' axis
    var rx = Math.cos(a)*x + Math.sin(a)*y;
    var ry = Math.cos(a)*y - Math.sin(a)*x;
    var rz = Math.cos(b)*z + Math.sin(b)*ry;
    ry = Math.cos(b)*ry - Math.sin(b)*z;
    // and project into 2D
    var px = rx * 2000 / (100-ry);
    var py = rz * 2000 / (100-ry);

    moveTo(px, py, function() {
      if (!once) sinrr(step+1);
    });
  }
  sinrr(0);
```

This renders the formula $\boxed{\text{sin r / r}}$, which looks a lot like a water droplet on a pond:

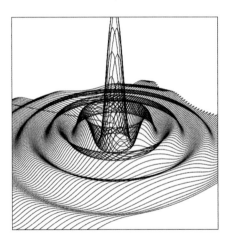

There's just one annoying problem. There's a line drawn from the center of the screen to the start of the drawing, and that's exactly what will happen if we try this on the plotter.

To solve this, I've added a parameter called $\boxed{\text{once}}$. When this is set to $\boxed{\text{true}}$, the $\boxed{\text{sinrr}}$ function will move the pen to the correct location but will do no more.

20. So let's draw this pattern on our plotter. Make sure there's no pen in the plotter.

21. Copy the following code on to the righthand side of the IDE and click *Upload*:

```
var StepperMotor = require("StepperMotor");
var motorx = new StepperMotor({
  pins:[B3,B4,B5,A6]
});
var motory = new StepperMotor({
  pins:[B10,B13,B14,B15]
});

function moveTo(x,y,callback) {
  var dx = x - motorx.getPosition();
  var dy = y - motory.getPosition();
  var d = Math.sqrt(dx*dx + dy*dy);
  var time = d * 1000 / motorx.stepsPerSec;
  motorx.moveTo(x, time);
  motory.moveTo(y, time, callback);
}

function sinrr(step, once) {
  // 'step' is going to keep increasing
  // make it 'scan' out in x and y
  var x = step % 100;
  var y = (step-x) / 100;
  if (y>=100) return;
  if (y&1) x = 100-x;
  // now center the coordinates on 0,0
  x -= 50;
  y -= 50;
  // Work out `r` - the radius
  //  - but add a bit to `r` to avoid a divide by 0 below
  var r = Math.sqrt(x*x + y*y) + 0.1;
  // Make 'z' a fun mathematical formula - (sin r)/r in this case
  var z = 100 * Math.sin(-r/2) / r;
  // now work out some 3D coordinates
  var a = 0.4; // rotation in 'y' axis
  var b = 0.5; // rotation in 'x' axis
  var rx = Math.cos(a)*x + Math.sin(a)*y;
  var ry = Math.cos(a)*y - Math.sin(a)*x;
  var rz = Math.cos(b)*z + Math.sin(b)*ry;
  ry = Math.cos(b)*ry - Math.sin(b)*z;
  // and project into 2D
  var px = rx * 2000 / (100-ry);
  var py = rz * 2000 / (100-ry);

  moveTo(px, py, function() {
    if (!once) sinrr(step+1);
  });
}

sinrr(0, true);
```

This will move the sled to the correct location for the start of the pattern.

22. Carefully insert your pen, and type `sinrr(0)` in the lefthand side of the IDE. This will start the drawing, and then it's just a matter of waiting! After some time, you should see a plot starting to appear:

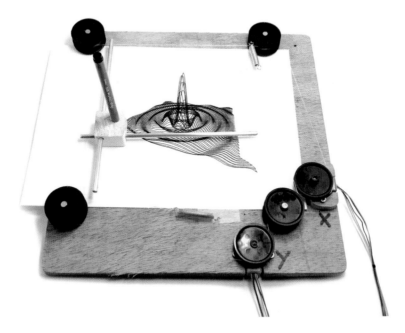

So now we've got a working plotter, and we can print mathematical formulae—but that could get boring quite quickly!

What if we could somehow connect our plotter to the internet so we could get it to draw images? Well, we can!

What Else Can I Do?

XY plotters are a great base for lots of projects; for example, many 3D printers consist of an XY plotter and then a table that moves up and down in the z-axis.

You could put a drag knife in the pen holder and use it for cutting shapes out of paper, or you could put a sharp pin in the pen holder and use it to scratch designs out on surfaces like copper.

You can also make you own automatic *zen garden* by fixing a magnet onto the plotter and then placing it underneath a bed of sand; you can make a steel ball draw patterns in the sand as it follows the magnet around!

Internet-Connected Plotter 18

While our plotter is connected by USB, we could send the data we want to output through the USB port. Unfortunately, we can't easily access the USB port from a web browser, so we'd have to make an application in a tool like Node.js.

It would be much better if we could use the Espruino to serve up a web page, and to have that web page handle the rendering, so that's what we're going to do!

First, we need to think of how we're going to draw our images. There are lots of things we could do, but one easy method is to scan from side to side with a sine wave much as we did for the water ripple in Chapter 17, and just to make the amplitiude of the sine wave larger when the image is darker. This will convert the input image (Figure 18-1) into a line drawing (Figure 18-2).

Figure 18-1 *Input image*

Use Image: [Choose file] make_logo.png

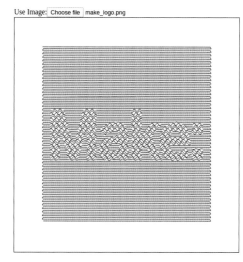

Figure 18-2 *Output line drawing*

If you want to play with this code on *https://jsfiddle.net*, all you need to do is add the following HTML:

```
Load image <input type="file" id="imageLoader" name="imageLoader"/><br/>
<canvas id="canv" style="border:1px solid black"></canvas>
```

And the following JavaScript:

```
// display the canvas
var canvas = document.getElementById("canv");
var imageWidth = 200;
var imageHeight = 100;
canvas.width = 500;
canvas.height = 500;
var ctx = canvas.getContext('2d');
var midx = canvas.width/2, midy = canvas.height/2;
var lastPos;

function moveTo(x,y,callback) {
  var pos = [x/4+midx, y/4+midy];
  if (lastPos)
    ctx.lineTo(pos[0], pos[1]);
  else
    ctx.moveTo(pos[0], pos[1]);
  lastPos = pos;
  setTimeout(callback,0);
}

function startDraw() {
  lastPos = undefined;
  ctx.clearRect(0,0,canvas.width,canvas.height);
  ctx.beginPath();
}
```

```
function endDraw() {
  ctx.stroke();
  ctx.closePath();
}

// Handle loading of the image
var imgData; // the raw RGBA image data
var imageLoader = document.getElementById('imageLoader');
imageLoader.addEventListener('change', function (e) {
  // This is called when you've chosen a file
  // First we read the file
  var reader = new FileReader();
  reader.onload = function(event) {
    // Then we load it into an image
    var img = new Image();
    img.onload = function() {
      // we draw that image on to our canvas
      ctx.drawImage(img, 0, 0, imageWidth, imageHeight);
      // read back the image data into an array
      imgData = ctx.getImageData(0, 0, imageWidth, imageHeight).data;
      // and finally we start converting it to a line drawing
      startDraw();
      scanImage();
     }
    img.src = event.target.result;
  }
  reader.readAsDataURL(e.target.files[0]);
}, false);

// Output the image as lines
function scanImage() {
  var step = 0;
  for (var y=0;y<imageHeight;y++) {
    for (var x=0;x<imageWidth;x++) {
      // get color from image - work out where (we want to zig-zag)
      var imagex = (y&1) ? x : imageWidth-x;
      var imagey = y;
      // the image is in RGBA format, so we take the average of
      // red, green and blue channels
      var col = (
        imgData[(imagey*imageWidth + imagex)*4] +
        imgData[(imagey*imageWidth + imagex)*4 + 1] +
        imgData[(imagey*imageWidth + imagex)*4 + 2]) / 3;
      // now work out where on the page to draw the line
      // and work out what
      var px = (imagex - imageWidth/2)*7.5;
      var py = (imagey - imageHeight/2)*15 +
            Math.sin(step)*(col-255)/15;
      step++;
      // and move to the location
      moveTo(px, py);
    }
  }
}
```

```
      endDraw();
   }
```

Click *Run* and then click *Choose File* on the web page that appears and select a square image from your PC.

The web page should now display an image like Figure 18-2. The function `scanImage` in the preceding code is the one that actually does the conversion.

All it does is scan backward and forward over the image, and makes the amplitude of the sine wave depend on the brightness of the picture at that point.

Now that we've got some working code, we need to get our plotter to serve up a web page containing it.

Experiment 39: Internet-Connected Plotter

You'll need:

- An Espruino WiFi, or an Espruino Pico and ESP8266 ESP01 module with patch wires

Follow these steps:

1. If you've got an Espruino Pico, wire it up as shown in Chapter 15. If you have an Espruino WiFi board you're ready to go!

2. Enter the following code in the righthand side of the Web IDE:

```
var WIFI_NAME = "";
var WIFI_KEY = "";
var wifi;

// Initialization for Espruino WiFi ONLY
function onInit() {
  wifi = require("EspruinoWiFi");
  wifi.connect(WIFI_NAME, { password : WIFI_KEY }, function(err) {
    if (err) {
      console.log("Connection error: "+err);
      return;
    }
    console.log("Connected!");
    wifi.getIP(function(err,ip) {
      console.log("IP address is http://"+ip.ip);
      createServer();
    });
  });
}

// Initialization for Espruino Pico + ESP8266 ONLY
function onInit() {
```

```
  Serial1.setup(115200, { tx: B6, rx : B7 });
  wifi = require("ESP8266WiFi_0v25").connect(Serial1, function(err) {
    if (err) throw err;
    console.log("Connecting to WiFi");
    wifi.connect(WIFI_NAME, WIFI_KEY, function(err) {
      if (err) {
        console.log("Connection error: "+err);
        return;
      }
      console.log("Connected!");
      wifi.getIP(function(err,ip) {
        console.log("IP address is http://"+ip.ip);
        createServer();
      });
    });
  });
}

// Create a web server on Port 80
function createServer() {
  var http = require("http");
  http.createServer(pageHandler).listen(80);
}

var mainPageContents = "Hello World";

// Called when a page is requested
function pageHandler(req, res) {
  var info = url.parse(req.url, true);
  //print(info);
  if (info.path == "/") {
    res.writeHead(200);
    res.end(mainPageContents);
  } else {
    console.log("Page "+info.path+" not found");
    res.writeHead(404);
    res.end("Not found");
  }
}
```

3. Make sure you change WIFI_NAME and WIFI_KEY to the correct ones for your network, and remove one of the onInit functions depending on whether you're using Espruino WiFi or Pico.

4. Click *Upload*, and when it's finished type **onInit()** on the lefthand side.

   If the WiFi module managed to connect you should see something like this:

```
|_| http://espruino.com
    Copyright 2016 G.Williams
>
=undefined
>onInit()
=undefined
Connected!
IP address is http://192.168.1.162
>
```

5. Now type 192.168.1.162 into the address bar of your web browser. You should see Hello World displayed there.

The next step is to change the web page that's served up to be the code that will trace the image we want.

Add to your previous code on the righthand side of the IDE until you get the following. The first section of code is identical to what was entered before.

This is a huge section of code as it contains the web page that we had previously tried in JSFiddle, as well as some code to send data to the Espruino, and the code for the Espruino itself!

```
var WIFI_NAME = "";
var WIFI_KEY = "";
var wifi;

// Initialization for Espruino WiFi ONLY
function onInit() {
  wifi = require("EspruinoWiFi");
  wifi.connect(WIFI_NAME, { password : WIFI_KEY }, function(err) {
    if (err) {
      console.log("Connection error: "+err);
      return;
    }
    console.log("Connected!");
    wifi.getIP(function(err,ip) {
      console.log("IP address is http://"+ip.ip);
      createServer();
    });
  });
}

// Initialization for Espruino Pico + ESP8266 ONLY
function onInit() {
  Serial1.setup(115200, { tx: B6, rx : B7 });
  wifi = require("ESP8266WiFi_0v25").connect(Serial1, function(err) {
    if (err) throw err;
    console.log("Connecting to WiFi");
    wifi.connect(WIFI_NAME, WIFI_KEY, function(err) {
      if (err) {
        console.log("Connection error: "+err);
```

```
      return;
    }
    console.log("Connected!");
    wifi.getIP(function(err,ip) {
      console.log("IP address is http://"+ip.ip);
      createServer();
    });
  });
});
}

// Create a web server on Port 80
function createServer() {
  var http = require("http");
  http.createServer(pageHandler).listen(80);
}


// Everything above here is the same
// ====================================

/* We're using an ES6 templated string here so
we can store the whole webpage verbatim, otherwise
we'd have to escape every single newline in the string. */
var mainPageContents = `<html>
<head><title>WiFi Plotter</title></head>
<body>
Load image <input type="file" id="imageLoader" name="imageLoader"/>
<br/><canvas id="canv" style="border:1px solid black"></canvas>
<script>
// display the canvas
var canvas = document.getElementById("canv");
var imageWidth = 200;
var imageHeight = 100;
canvas.width = 500;
canvas.height = 500;
var ctx = canvas.getContext('2d');
var midx = canvas.width/2, midy = canvas.height/2;
var lastPos;
// list of x,y,x,y points for the plotter
var plotPoints = [];

function sendToEspruino() {
  var points = [];
  var pointCount = 40; // amount of plot data to send at once
  if (pointCount>plotPoints.length)
    pointCount = plotPoints.length;
  if (pointCount==0) {
    console.log("Done!");
    return;
  }
```

```
  // Get the data to send
  points = plotPoints.slice(0, pointCount);
  // send the data to Espruino
  httpRequest = new XMLHttpRequest();
  httpRequest.open('POST', 'push?pts='+points.join(","), true);
  httpRequest.timeout = 1000; // timeout in milliseconds
  httpRequest.onreadystatechange = function(){
    if (httpRequest.readyState === XMLHttpRequest.DONE) {
      if (httpRequest.status === 200) {
        var response = httpRequest.responseText;
        console.log("Got response "+response);
        if (response=="busy") {
          // try again after a delay
          setTimeout(function() {
            sendToEspruino();
          }, 2000);
        } else {
          // We sent it! Delete these points from our list
          plotPoints.splice(0, pointCount);
          // Wait a little and carry on with sending
          setTimeout(function() {
            sendToEspruino();
          }, 500);
        }
      } else {
        console.log('There was a problem with the request.');
        // Try again after a delay
        setTimeout(function() {
          sendToEspruino();
        }, 2000);
      }
    }
  };
  console.log("Sending "+pointCount+" points");
  httpRequest.send(null);
}

function moveTo(x,y,callback) {
  plotPoints.push(x,y);
  var pos = [x/4+midx, y/4+midy];
  if (lastPos)
    ctx.lineTo(pos[0], pos[1]);
  else
    ctx.moveTo(pos[0], pos[1]);
  lastPos = pos;
  setTimeout(callback,0);
}

function startDraw() {
  lastPos = undefined;
  ctx.clearRect(0,0,canvas.width,canvas.height);
```

```
  ctx.beginPath();
}

function endDraw() {
  ctx.stroke();
  ctx.closePath();

  sendToEspruino();
}

// Handle loading of the image
var imgData; // the raw RGBA image data
var imageLoader = document.getElementById('imageLoader');
imageLoader.addEventListener('change', function (e) {
  // This is called when you've chosen a file
  // First we read the file
  var reader = new FileReader();
  reader.onload = function(event) {
    // Then we load it into an image
    var img = new Image();
    img.onload = function() {
      // we draw that image on to our canvas
      ctx.drawImage(img, 0, 0, imageWidth, imageHeight);
      // read back the image data into an array
      imgData = ctx.getImageData(0, 0, imageWidth, imageHeight).data;
      // and finally we start converting it to a line drawing
      startDraw();
      scanImage();
     }
    img.src = event.target.result;
  }
  reader.readAsDataURL(e.target.files[0]);
}, false);

// Output the image as lines
function scanImage() {
  var step = 0;
  for (var y=0;y<imageHeight;y++) {
    for (var x=0;x<imageWidth;x++) {
      // get color from image - work out where (we want to zig-zag)
      var imagex = (y&1) ? x : imageWidth-x;
      var imagey = y;
      // the image is in RGBA format, so we take the average of
      // red, green and blue channels
      var col = (
        imgData[(imagey*imageWidth + imagex)*4] +
        imgData[(imagey*imageWidth + imagex)*4 + 1] +
        imgData[(imagey*imageWidth + imagex)*4 + 2]) / 3;
      // now work out where on the page to draw the line
      // and work out what
      var px = (imagex - imageWidth/2)*7.5;
```

```
      var py = (imagey - imageHeight/2)*15 +
            Math.sin(step)*(col-255)/15;
      step++;
      // and move to the location
      moveTo(px, py);
    }
  }
}
endDraw();
}
</script>
</body>`;

// Called when a page is requested
function pageHandler(req, res) {
  var info = url.parse(req.url, true);
  if (info.pathname == "/") {
    res.writeHead(200);
    res.end(mainPageContents);
  } else if (info.pathname == "/push") {
    // we got
    res.writeHead(200);
    var accepted = queueMove(info.query.pts.split(","));
    res.end(accepted ? "ok" : "busy");
  } else {
    console.log("Page "+info.path+" not found");
    res.writeHead(404);
    res.end("Not found");
  }
}

// Stepper motor handling
var StepperMotor = require("StepperMotor");
var motorx = new StepperMotor({
  pins:[B3,B4,B5,A6]
});
var motory = new StepperMotor({
  pins:[B10,B13,B14,B15]
});

function moveTo(x,y,callback) {
  var dx = x - motorx.getPosition();
  var dy = y - motory.getPosition();
  var d = Math.sqrt(dx*dx + dy*dy);
  var time = d * 1000 / motorx.stepsPerSec;
  motorx.moveTo(x, time);
  motory.moveTo(y, time, callback);
}

var busy = false;
var nextPositions = [];
function queueMove(positions) {
```

```
if (nextPositions.length>40) {
  // we already have a lot of positions queued
  console.log("Rejected positions");
  return false; // don't take any more
} else {
  // add something else onto the queue
  console.log("Queued "+(positions.length/2)+" positions");
  for (var i=0;i<positions.length;i+=2)
    nextPositions.push([positions[i],positions[i+1]]);
}
// not busy moving - start!
if (!busy) {
  busy = true;
  moveFinished();
}
// return true to show the position was accepted
return true;
}
function moveFinished() {
  // we just finished - see if there's anything else
  if (nextPositions.length>0) {
    // there is - get a new position off our queue
    var nextPos = nextPositions.shift();
    // go there
    moveTo(nextPos[0], nextPos[1], moveFinished);
  } else {
    // no - we're no longer busy
    busy = false;
  }
}
```

6. Now click *Upload* again and type $\boxed{\text{onInit()}}$ on the lefthand side of the IDE.

7. You should see an IP address displayed again (usually this is the same as it was before). Enter the address in the web browser again, and you should see a web page like this:

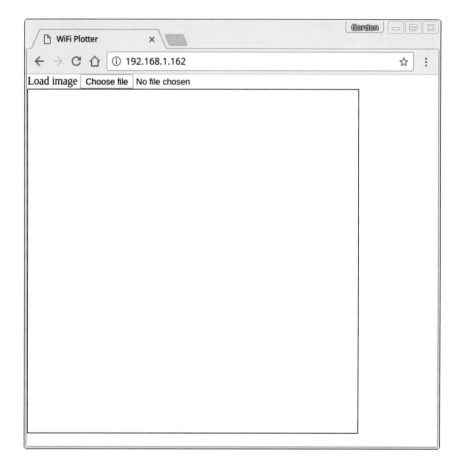

8. If you choose an image to upload, it will immediately be scanned, and drawing will start! Eventually you should end up with a drawing like this:

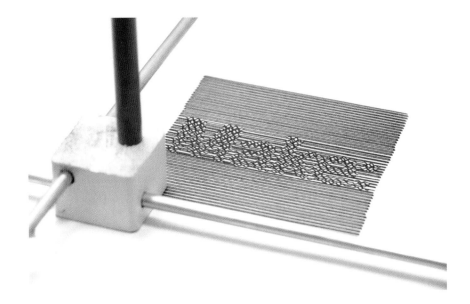

So What's Happening Here?

Espruino is serving up a web page, much like JSFiddle did. When you upload an image to the web page, the JavaScript running on your PC will convert that to a series of 2D points that make up a long line. They're saved in a variable called `plotPoints`.

Then, the function `sendToEspruino` is called on your PC, and this requests a web page of the form where `x1,y1` is the first point, `x2,y2` is the second, and so on:

```
http://ip.address/push/
pts=x1,y1,x2,y2,x3,y3,x4,y4,...
```

The Espruino will then take that information and add it to a queue called `nextPositions`, which is used to call `moveTo` as we used in the last chapter.

Espruino doesn't have enough memory to hold all the data we're sending it, so if it thinks it has enough data to keep it busy it'll reply `busy` to the web request, and the web page on the PC will have to try to send the data again in a few seconds.

There are other ways of transferring this data, for example, with a web technology called Web-Sockets. However, this is nice and easy, uses what we've already learned, and works well enough for us!

What Else Can I Do?

Now that you've got a plotter you can control from the internet you could do all kinds of things.

1. You could modify the `scanImage()` function to create different patterns based on the source image.

2. The web page could automatically download images from a source like a photo-sharing website and could plot those.

3. The website could use the `getUserMedia` API to use the webcam to take a picture to use as the basis for the plotting.

4. You could modify the sled on the plotter to take a servo that could raise and lower a pen. It could even hold a piece of cloth and then you could use the plotter to draw on a whiteboard, and then erase itself!

PART VI

Conclusion

Many of the machines in this book have been very basic, requiring very little skill or care during assembly. While you can get better results from less wobbly machinery, I hope that if nothing else this book has shown you that you don't need amazing skills, tools, or expensive components to make fun devices.

You've learned how to control motors, use sensors, and then how to use a microcontroller to combine everything to get the right result from very simple machinery.

But we haven't even covered some of the most exciting parts of microcontroller development: adding displays, interesting sensors, and making multiple devices work together.

So What Now?   13

If you fancy exploring microcontrollers at a very low level, there's a section on writing assembly code in Appendix B. Once you're done with this book there's a wealth of information online. There are loads of tutorials on the Espruino website (*http://espruino.com*), including a handy search tool; for instance, try searching for "Humidity" to get a list of sensors with documentation and libraries for Espruino along with information on where to get them.

If you're interested in learning more about electronics so you can connect more complicated things to Espruino, there are great books like *Make: Electronics* by Charles Platt or *Making Things Talk* by Tom Igoe. If you want to learn how the Espruino JavaScript interpreter works, then I wholeheartedly recommend the "dragon book," *Compilers: Principles, Techniques, and Tools* by Aho, Sethi, and Ullman.

There are also some amazing websites, like Instructables (*http://www.instructables.com*) and Hackster (*https://www.hackster.io*), that are full of ideas for all kinds of projects. There are hundreds of great project ideas on YouTube too, or check out Make: Magazine's website (*http://makezine.com*) for some fantastic curated projects and articles.

But most of all, get stuck in and have fun!

Parts and Materials

A

What follows is a list of parts you'll need for this book, and where to get them. I've grouped together the parts you'll need for multiple sections first. It's also worth checking out the *Making Things Smart* GitHub page (*https://github.com/espruino/making-things-smart*) as it contains links to common online sources of some of the parts listed here.

Common Parts

General

- Cellophane tape
- Black masking/electrical tape
- Elastic bands
- Paperclips
- Pencils, pens, felt-tip pens
- USB power pack
- USB Type A extension lead

Espruino Boards

You can get Espruino boards from several different distributors. See the Espruino order page (*http://www.espruino.com/Order*) for an up-to-date list.

Mainly we use a Espruino Pico in this book, but toward the end we also use Espruino WiFi (or an Espruino Pico and an ESP8266 module) and Puck.js.

Espruino is open source so it has been ported to other boards as well. For most of the projects here it is possible to use Espruino on those boards, but it's possible you may hit a few bumps along the way! See the Espruino website (*http://www.espruino.com/Other +Boards*) for an up-to-date list of supported boards.

Breadboard

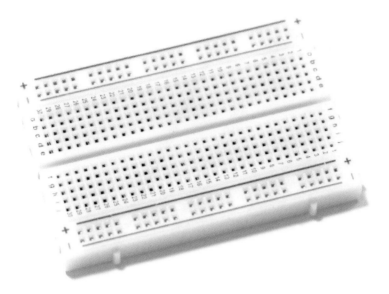

Lots of the projects here use breadboard to avoid soldering. Pretty much any hobby electrical store will sell breadboard, or you might find that buying an Electronics Starter Kit online is a cheap way to get breadboard along with a lot of the other components mentioned here.

Patch Wires/Jumper Leads

Patch wires are just wires to connect electronic components up with, usually on bread-board. Normal patch wires are pointy at both ends (we call each end *male*). While these can be nice, all you really need is some of the solid core wire that's also on this list.

However, for the final chapter we use male-to-female jumper wires. These have a socket on one end, and a pin on the other, and are really useful especially if you plan to avoid solder-ing.

If you bought an electronics or Arduino starter kit then you may well have some, however you can find them online by searching for `dupont male female`.

You can also get female-to-female leads, which aren't used in this book but are still extremely useful (and can be combined with the male-to-male leads to make a male-to-female lead).

Resistors, Capacitors, LEDs, Light-Dependent Resistors

You can buy these components separately (but a few different values of resistor will be needed for your projects). In most cases it will be cheaper to buy an Electronics Starter Kit. These can be bought online for around $30 and will contain everything you need, as well as a few of the extra parts in this appendix!

Neodynium Magnets

These are really strong, compact magnets. You can buy them online from places such as Amazon or eBay, or electrical stores such as Radio Shack in the US or Maplin in the UK will stock them.

If the magnets don't need to be circular then you can also find them by dismantling old computer hard disk drives (you won't find them in modern SSDs though!)

Single Core Wire

Single core wire isn't great for wiring that needs to move around, but it is handy for wiring up breadboards if you don't have any patch wires. A wire with a conductor diameter of between 0.6mm and 0.8mm (20 to 23 AWG) is perfect, but you want one with insulation on it as well.

You can get wire from online electrical stores very easily, or can find it in the kind of networking cable that's designed for permanent installations.

Wood

For various experiments you'll need:

- A block of wood, roughly 10cm×5cm×2cm

- A block of wood, roughly 7cm×7cm×7cm

- 1cm-thick square sheet of plywood, at least 30cm×30cm

- A 30mm×30mm×30mm cube of solid wood or plastic

Like most things in this book, these measurements aren't exact. If you have any offcuts lying around or find something in a dumpster, chances are it'll be about right!

Motors Section

General

- Nails

- Screws

- Wine bottle cork

- White sticky labels

- A flashlight

A Source of DC Power (Around 6–12v)

For the motors in this section, you'll need a power source that'll provide quite a few amps of current. There are a few potential options here:

- PP9 9v battery, PJ996 6v battery

- 4 to 8 AA batteries in a battery holder

- Model racing car/airplane batteries (make sure you get the correct voltage and be careful if you have batteries without protection circuitry, as they can easily melt things if shorted!)

- 9.6v or 12v power tool batteries as long as you can safely connect to them

- Bench power supply

You'll be able to find one or another of these at a local electronics store or online.

L293D Motor Driver IC

You can buy L293D chips online from most electronics retailers. Just make sure you buy the *plastic dual-in-line package* with the big pins rather than the *surface mount* variants (that won't fit in breadboard).

If you don't want to do the wiring up you can get many other general-purpose motor drivers, some of which come on a handy board with screw terminals. Search online for `h bridge motor driver`.

Brushless Fan

See Chapter 6 where finding a fan is discussed.

P36NF06L FET

The P36NF06L FET is a good general-purpose power transistor that will work well off the relatively low (3.3v) voltages from the Espruino Pico. You can find them online from many electronics retailers, but any low voltage FET will work just as well for us.

However, if you have the L293D (or H-bridge) motor driver IC then you can use one channel of that instead. If you have the control board from a stepper motor that is used in Chapter 17 (see the following section) then you can use that as well.

You could also use an NPN transistor with a resistor if you have them handy, but wiring transistors up is out of the scope of this book!

Electromechanics Section

General

- Chopsticks
- Thick noncorrugated cardboard
- Thin string (kite string is perfect)
- A Pringles can or whisky tube
- A sturdy cardboard box, 30cm on one side

Servo Motors and Extension Wire

You can buy servo motors from most model shops, however the motors we're using for the projects in this book are *9g* servos (referring to their weight of 9 grams). You may find it easiest to purchase them online.

These servos are by no means amazing, but they are very cheap, and a great base for projects. However, if you do decide you need to improve what you've made, you can get bigger, more expensive servos that can be connected in exactly the same way.

There are a few manufacturers, but Feetech/Fitech and Towerpro appear to be good makes of 9g servos. Continuous rotation 9g servos are harder to come by, but Fitech make some with the model number FS90R, which is easy to find on eBay once you've scrolled past all the results for Stihl grass trimmers.

Finally, servo motor extension wire is needed for the plotter, and you can also get this online or from a hobby shop. Failing that, if you have bought male-to-female patch wire (see the previous discussion) then you can use that, as long as you're sure you have kept the wires in the right order!

Lobster Bands

Any short, wide elastic band will work to give the little robot wheels some grip, but I've found lobster bands to be perfect. These are the elastic bands that get put on lobsters to stop them from pinching things after they've been caught.

The easiest and most enjoyable way to get lobster bands is to buy (and then eat) a lobster, but it's not the cheapest. You could probably ask at a local restaurant, or you can buy a bag of thousands of lobster bands online for under $10.

A Small Corkboard (Roughly 30cm×40cm)

You can get corkboards from many stationery shops, or home improvement shops like Ikea can also be a good source.

Threaded Rod and Nut

Threaded rod can usually be bought from DIY/home improvement shops or builders' merchants.

M6 or M8 thread (6mm or 8mm diameter) is perfect. But don't forget that you'll also want a nut and some penny (large diameter) washers to go with it!

Communication Section

A Headphone Lead with a 3.5mm Jack Plug

You'll need to be willing to destroy this by removing the headphones from it!

Try to get one that has a single cable (a circular cross-section rather than a figure of 8).

IR Receiver (HX1838, VS1838, TSOP348, or TSOP344) and Remote Control

If you bought an Arduino starter kit you may well find one of these inside, otherwise you can usually find them by searching online for `arduino infrared kit`.

Failing that, they're usually inside old consumer electronics as well—so if you have an old VCR or DVD player that you're throwing out you may be able to rescue the IR receiver from it!

A 315Mhz (USA) or 433Mhz (Europe) Radio Transmitter/Receiver

You can usually get these online on Amazon or eBay. Just search for `315Mhz` (or `433Mhz`) `rf receiver` and look for something like this:

The transmitters are generally square little boards with a large silver component on them. You can almost always buy transmitters and receivers as a pair, which would be a great idea for Chapter 14.

ESP8266 ESP01

ESP8266s come in many different types of module—ESP12, ESP07, etc. Unless you have some kind of breakout board you really need the ESP01, as most of the others are difficult to connect to.

Just searching online for `ESP8266 ESP01` will find you loads of suppliers, but sadly this is a part that you're unlikely to find in a local shop.

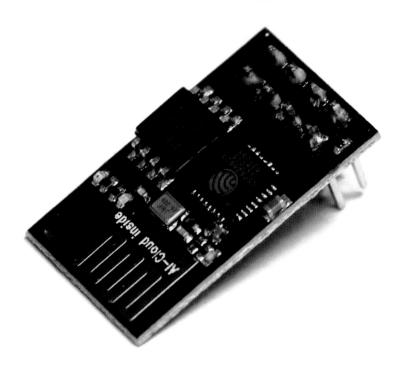

Putting It All Together

2x Springs

You just want springs that are designed to be pulled apart, with relatively little force. Springs from ballpoint pens will work, but if you are able to scavenge proper springs from something like an old CD-ROM/DVD drive then they'd be perfect!

Smooth Aluminum Bar and 30mm×30mm×30mm Cube of Solid Wood or Plastic

The bar is to be used for the X and Y crossbars of the plotter. The cube of wood will have holes drilled in it so that it can slide smoothly over the bar.

For the plotter in this book I used a 5mm diameter aluminum welding rod and a cube of MDF wood because that's what I had handy. The rod did have numbers stamped on it, but I sanded them off.

In reality you could use anything here, even a plastic rod. You just need to be confident that the rod will be able to slide smoothly through a hole in the cube.

10×40mm Diameter Pulleys

These pulleys can be slightly harder to find, but you should be able to get them in a model shop. Failing that, you can often find them advertised online as a Plastic Model Pulley, or if you have access to a 3D printer you could print some yourself.

The pulleys don't need to be 40mm, but given the relatively low torque of the stepper motors used I wouldn't advise using a larger diameter.

You also need to find four circular nails that will fit inside the pulleys. The pulleys pictured here have a 3mm hole, which makes finding nails that fit inside nicely really easy!

5M of Fishing Line

Any kind of thin nylon line should work well. However, fishing line is relatively easy to find, and can be bought in outdoors shops or even some large supermarkets.

Two Small Geared Stepper Motors

These stepper motors are very common, cheap Chinese motors. Probably the easiest way to find them is to look on eBay or Amazon. You can either search for `5v stepper` or can search for the part number, which is `28BYJ-48`.

It really helps to get them with the ULN2003 motor driver board as well, as it will have a connector that fits directly on to the motor. When ordered together they can be purchased for as little as $3.

Common Espruino Commands and Variables B

What follows is a quick rundown of the most common Espruino commands. For a full list, check out the Espruino reference page (*http://www.espruino.com/Reference*).

 Where parameters are surrounded by `[` and `]`, they're optional and can be left out.

print(text) or console.log(text)

This prints the given text or variable to the console. `console.log` is the standard Java-Script way of printing, and `print` is provided for convenience. Both commands are identical.

For example, `a="Hello World";print(a)` will write `Hello World` to the console, as would `print("Hello World")`.

You can also supply multiple arguments that will all be printed, separated by spaces.

LED1 and LED2

These variables are the pins that are connected to the onboard LEDs. For example, writing `digitalWrite(LED1, 1)` will light what is usually the red LED.

BTN1 or BTN

This variable is the pin that is connected to the onboard button. For example, writing `digitalRead(BTN1)` will return 1 if the button is pressed.

digitalWrite(pin[s], value)

If the pin state has not previously been set with `pinMode`, this sets the GPIO pin to be a digital output, and then outputs either a 1 (3.3v) or a 0 (0v) to the pin depending on the value.

If you supply more than one pin in an array, the value is treated as a binary number. For example, `digitalWrite([LED1,LED2], 0b10)` will turn `LED1` on and `LED2` off.

This is different from an Arduino in that you don't have to use `pinMode` on Espruino if you don't want to.

digitalRead(pin[s])

If the pin state has not previously been set with `pinMode`, this sets the GPIO pin to be a digital input, and then reads the value on the input, returning either a 1 or a 0.

If you supply more than one pin in an array, the returned value is a binary number. For example `digitalRead([B3,B4])` will return `2` if `B3` is 1 and `B4` is 0.

This is different from an Arduino in that you don't have to use `pinMode` on Espruino if you don't want to.

Pin.read(), Pin.write(value), Pin.set(), and Pin.reset()

These commands work similarly to `digitalWrite` and `digitalRead`, but they are available on each pin variable. Sometimes it's just more concise to write `LED1.set()` than `digitalWrite(LED1, 1)`.

analogWrite(pin, value[, options])

`value` can be any floating-point value between 0 and 1.

This writes an analog voltage to the given pin. By default this uses specific peripherals that aren't available on all pins (look for `DAC` or `PWM` in the pin chart for the device you're using).

On devices with a DAC (like the original Espruino board), an actual analog voltage is output on the pin, ranging from 0 to 3.3v depending on `value`. On most devices there isn't a DAC, in which case this will use pulse width modulation of the digital output to output (on average) an analog value. In this case, `options` can be an object with a `freq` element to specify the frequency (for example, `analogWrite(B3, 0.5, { freq: 100 });`).

On some pins there is no DAC or PWM, in which case Espruino can do PWM in software (at frequencies below 10KHz). For this you have to specify `soft:true` in the object: `analogWrite(LED1, 0.1, { freq: 100, soft:true });`.

analogRead(pin)

This reads an analog value from the given pin. The functionality is only available on some pins. Look for pins on the pin chart that have `ADC` next to them.

The value returned is a floating-point value between 0 and 1 (this is different than Arduino, which returns an integer between 0 and 1023). A value of 1 corresponds to the voltage that the microcontroller is powered off, which is usually 3.3v. You can check this with the command `E.getAnalogVRef()` though.

Be careful: most pins that have analog inputs only accept voltages between 0 and 3.3v (as opposed to a lot of other pins that will often go up to 5v), so try not to connect the pin to anything where the voltage might go too high.

digitalPulse(pin, polarity, time[s])

This sends an accurately timed pulse to the given pin. For example, `digitalPulse(LED1, 1, 100)` will turn `LED1` on (because polarity was 1) for 100ms, and will then turn it off.

You can also supply an array. For example, `digitalPulse(LED1, 0, [100,50,25])` will turn `LED1` off for 100ms (because polarity was 0), then on for 50ms, then off for 25ms, and finally will leave it on.

`digitalPulse` uses interrupts to send accurate pulses, which means it returns immediately, even while the pulses are being generated in the background. This can be really powerful, but it can also trip you up very easily! If you don't want this, call `digitalPulse(A0,1,0)` immediately afterwards. This will cause `digitalPulse` to return only after all pulses have been sent.

pinMode(pin, mode)

This sets the mode of the pin (whether it is an input or output).

Common calls are:

- `pinMode(pin, 'input')` sets the pin to be a normal input.
- `pinMode(pin, 'input_pulldown')` sets the pin to be an input, but with an internal pull-down resistor (to 0v) enabled.
- `pinMode(pin, 'input_pullup')` sets the pin to be an input, but with an internal pull-up resistor (to 3.3v) enabled.
- `pinMode(pin, 'output')` sets the pin to be a normal output, which will output either 0v or 3.3v.
- `pinMode(pin, 'opendrain')` sets the pin to be an output, which will output 0v when 0 is written, but will leave the output disconnected when 1 is written.

Unlike Arduino you don't have to call this before `digitalWrite`/etc. They will automatically set the correct state unless you have explicitly called `pinMode` beforehand.

You can use `getPinMode(pin)` to check the current state of a pin.

reset()

This resets Espruino, removing all code and setting all pins to their default values. It's usually called by the Web IDE before uploading your code, to ensure that it behaves the same way each time.

save()

This saves *the current state of Espruino* to read-only memory, which will then be loaded at power-on.

Afterwards, Espruino resumes from the saved state (as if you'd called `load()` or powered Espruino on).

 This doesn't actually save the code you wrote on the righthand side of the web IDE. The code you wrote was executed as you uploaded it, and the result is saved. For example, if you wrote `var a = E.getTemperature();`, `a` *will be set to the temperature of Espruino when you uploaded the code, rather than when the code was loaded. If you wrote* `setTimeout(function() { console.log("Hello");}, 10000/* 10 seconds */);` *uploaded, and then saved nine seconds later, you would find that after loading* `Hello` *was printed only one second after boot. If you want to actually execute code at startup, check out* `onInit` *below.*

load()

This loads the state of Espruino from read-only memory (that was previously saved with `save()`). It's a bit like unplugging and replugging the Espruino board.

onInit()

This is a special function you can write that is called when Espruino starts up after code is saved with `save()`, either when the board is powered on, or after `load()` or `save()`.

Calling `E.on('init', function() { ... my code ... });` has the same effect, except you can call `E.on` multiple times with different functions to queue up. There can only ever be one `onInit` function.

Espruino Assembler

C

As mentioned in Chapter 2, Espruino's Web IDE has a built-in assembler.

Assembly language is a textual representation of the actual instructions that the computer executes. The ARM microcontroller in the Espruino boards executes what is called *Thumb* code. This is a special cut-down version of ARM assembly code that is more limited, but takes up half the code space per instruction (16 bits versus 32 bits).

A very simple Thumb function might look like this:

```
mov  r0, #42
bx   lr
```

This function consists of two 16-bit instructions (so it is just 4 bytes long). The first instruction loads register `r0` with the number `42`, and the second instruction returns from the function. If you didn't have this, then the microcontroller would just keep executing operations past the end of the function, which would probably result in a crash!

So How Do We Run This Code?

To use the code, we have to wrap it up in something that looks like valid JavaScript code. The Web IDE detects any function calls to `E.asm`, intercepts them, and replaces them with the assembled code.

Enter the following on the righthand side of the IDE and click *Upload*:

```
var fortytwo = E.asm("int()",
"mov  r0, #42",
"bx   lr");

print(fortytwo());
```

You should now have `42` printed…

But what does all this mean?

- On the first line `var fortytwo = E.asm("int()",`, we're creating a variable called `fortytwo`, which will hold our function. `E.asm` tells the Web IDE to assemble what follows, and `"int(void)"` tells the IDE that the assembler is to be a function that takes no variables (`()`) and returns one integer (`int`). While that's not part of the assembler, Espruino needs to know so that it can create a valid JavaScript function that you can call.

- `"mov r0, #42"` is the first assembler instruction. The `mov` bit means `move`. `r0` is the register we'll move to (more on registers in a bit), `#` means we're going to specify an actual number (a literal) and `42` is our number.

- Finally, `bx lr` stands for `Branch with eXchange, Link Register`. So what does that mean? Well, when you call a function, the ARM sets `lr`, the *Link Register*, to the address it was just about to execute at the time of the call. When you do `bx lr`, it jumps back to the address in `lr`, and keeps executing the function that called this one where it left off. It's just like a `return` statement in a JavaScript function.

Registers

ARM cores have 16 registers that can be used by each instruction; each one of them is 32 bits wide. While only the last register (r15) has a special function, several other registers are reserved by convention.

r15 (PC)
> This is the Program Counter; it is the actual address in memory of the instruction that's being loaded at that time. Because the ARM core is made to be simple, but is pipelined, the Program Counter doesn't actually point to the instruction that's being executed, but to the instruction that's being loaded. This means that it's actually pointing *two whole instructions* ahead of the currently executing function.

r14 (LR)
> This is the Link Register; when executing a function, this register contains the address of the caller. If you want to be able to recurse, you have to manually save this register onto the stack or it'll just get overwritten.

r13 (SP)
> This is the Stack Pointer; it points to the top of the stack. On ARM the stack usually grows downward in memory, so as more data gets added to the stack the value of SP gets smaller, and it rises as things are taken off the stack.

r0 to r3

> These are treated as normal registers, but when calling functions they're used for the first four arguments (and r0 is used for the return value). Everything else is stored on the stack.

There are also some hidden registers that get swapped in when you're called from an interrupt, but we won't go into those.

Finally, there are the condition flags. These are just a handful of bits. They aren't directly accessible as a register, but they can be used to determine whether an instruction (for example, a jump) is executed or not. When you execute a command such as CMP (Compare), the condition flags are updated depending on the result:

- Z—is the result zero?

- N—is the result negative?

- C—did the last math operation result in a carry?

- V—did the last math operation result in a overflow?

You can then have a command such as BZ, which means: Branch *only* if the condition flags say the last result was zero.

Instructions

As we hinted at previously, Thumb assembler is a very cut-down version of ARM assembler. ARM assembler is a classic RISC instruction set. Each 32-bit instruction is formatted as shown in Table C-1 (there are some exceptions).

Table C-1 *An average ARM instruction*

Bits	Usage
31-28	Condition flags
27-20	Type of operation
19-16	Register 1
15-12	Register 2
11-0	Extra data (sometimes Register 3)

This is really nice. It means that the hardware that decodes the instruction can be really simple, as the data for Register 1 (for instance) is always in the same place.

Unfortunately it also means that there's some wasted space because not all operands make sense for all kinds of instructions, and on a small embedded microcontroller space is at a real premium.

So ARM came up with Thumb. You can think of a Thumb microcontroller as being a normal ARM microcontroller that takes 32-bit instructions with a decoder stuck on the front of it, which takes the smaller more complicated 16-bit instruction, and turns it into an equivalent 32-bit instruction.

In practice this works out well, but it does mean that some instructions that make perfect sense in the ARM world just aren't available on Thumb. For instance, in ARM assembler you have access to all 16 registers for basically every instruction, but in Thumb you only have access to 8 registers for most instructions, with only a few capable of accessing the full 16.

Table C-2 is a very brief list of common ARM Thumb instructions. There are plenty more, and it's best to search online for one of ARM's User Guides to get a full list.

Table C-2 *Common ARM instructions*

Instruction	Description
MOV rD, rS	Set rD to rS
MOV rD, #val	Set rD to the literal val
ADD rD, rA, rB	Set rD to rA + rB
ADD rD, rA, #val	Set rD to rA + val
SUB rD, rA, rB	Set rD to rA - rB
SUB rD, rA, #val	Set rD to rA - val
NOP	Don't do anything! A *no-op*
STR rS, [rA,rB]	Store rS in memory at the address made by adding rA to rB
LDR rS, [rA,rB]	Load rS from the memory address made by adding rA to rB
PUSH {r1...r7,lr}	Push the given registers onto the stack
POP {r1...r7,lr}	Pop the given registers off the stack
B address	Start executing from the given address
CMP rA, rB	Compare registers rA and rB

Instruction	Description
BNE address	Start executing from the given address, but only if CMP's arguments weren't equal (Compare flags != 0)
BEQ address	Start executing from the given address, but only if CMP's arguments were equal (Compare flags = 0)
BGT address	Start executing from the given address, but only if CMP's left argument was greater than the right (Compare flags > 0)
BGE address	Start executing from the given address, but only if CMP's left argument was greater than or equal to the right (Compare flags >= 0)
BLT address	Start executing from the given address, but only if CMP's left argument was less than the right (Compare flags <= 0)
BLE address	Start executing from the given address, but only if CMP's left argument was less than or equal to the right (Compare flags <= 0)

Getting More Complex

So now that we have a rough idea what's happening and what instructions are available, let's try to write some assembler that's a bit more complex.

First, let's just add two numbers together. As mentioned in "Registers", the first four arguments are passed in registers r0 to r3, and the return value is put in r0. This makes a lot of functions nice and easy. The first argument is in r0, the second is in r1, and we just want to put the result back in r0.

Note that we've changed the first argument of E.asm from "int()" to "int(int,int)". This tells Espruino that the function now takes two arguments.

```
var add = E.asm("int(int,int)",
"add     r0, r0, r1",
"bx  lr");

print(add(1, 2));
```

Next, we might want to do some kind of loop, which we can do using bgt (branch if greater than):

```
var add = E.asm("int(int,int)",
"loop:",                 // label of the start of our loop
"add     r0, r0, r1",    // add r1 to r0
"sub     r1, #1",        // subtract 1 from r1
"bgt     loop",          // if r1 was greater than or equal to 0
after                    // 'sub', go back to the beginning
"bx  lr");
```

```
for (var i=0;i<10;i++)
  print(add(0, i));
```

This is roughly equivalent to the JavaScript code:

```
function add(r0, r1) {
  do {
    r0 = r0 + r1;
    r1--;
  } while (r1>0);
  return r0;
}
```

Now that you've got loops you've got the basics and are ready to start writing more complex code in assembler. Just remember, you're running code at the lowest level, so if you make an infinite loop by accident you won't be able to Ctrl-C your way out of it. It's time to unplug and re-plug the board to reset it!

Index

C

About the Author

Gordon Williams is an entrepreneur and inventor living near Oxford, UK. He grew up writing software and playing with electronics as a young child and went on to study Computer Science at Cambridge University, England.

Gordon worked for a variety of technology companies, specializing in 3D graphics and compiler design in a variety of languages. He's been working on the Espruino JavaScript interpreter since 2012, single-handedly developing and launching three successful crowd-funding campaigns and four different Espruino devices.

Gordon now works full-time developing Espruino, supporting the amazing community of Espruino users worldwide and attending and speaking at events within the JavaScript, embedded software, and Maker communities all over the world.

Colophon

The cover images were photographed by Gordon Williams. The cover font is Benton Sans Bold. The text font is Adobe Myriad Pro; the heading font is Benton Sans Bold; and the code font is Dalton Maag's Ubuntu Mono.